GAME THEORY

GAME
THEORY

Hot Science is a series exploring the cutting edge of science and technology. With topics from big data to rewilding, dark matter to gene editing, these are books for popular science readers who like to go that little bit deeper ...

AVAILABLE NOW AND COMING SOON:

Destination Mars:
The Story of Our Quest to Conquer the Red Planet

Big Data:
How the Information Revolution
is Transforming Our Lives

Gravitational Waves:
How Einstein's Spacetime Ripples Reveal the Secrets
of the Universe

The Graphene Revolution:
The Weird Science of the Ultrathin

CERN and the Higgs Boson:
The Global Quest for the Building Blocks of Reality

Cosmic Impact:
Understanding the Threat to Earth from Asteroids
and Comets

Artificial Intelligence:
Modern Magic or Dangerous Future?

Astrobiology:
The Search for Life Elsewhere in the Universe

Hot Science series editor: Brian Clegg

GAME
THEORY

UNDERSTANDING THE
MATHEMATICS OF LIFE

Brian Clegg

ICON

Published in the UK and USA in 2022
by Icon Books Ltd, Omnibus Business Centre,
39–41 North Road, London N7 9DP
email: info@iconbooks.com
www.iconbooks.com

Sold in the UK, Europe and Asia
by Faber & Faber Ltd, Bloomsbury House,
74–77 Great Russell Street,
London WC1B 3DA or their agents

Distributed in the UK, Europe and Asia
by Grantham Book Services,
Trent Road, Grantham NG31 7XQ

Distributed in the USA
by Publishers Group West,
1700 Fourth Street, Berkeley, CA 94710

Distributed in Australia and New Zealand
by Allen & Unwin Pty Ltd,
PO Box 8500, 83 Alexander Street,
Crows Nest, NSW 2065

Distributed in South Africa
by Jonathan Ball, Office B4, The District,
41 Sir Lowry Road, Woodstock 7925

Distributed in India by Penguin Books India,
7th Floor, Infinity Tower – C, DLF Cyber City,
Gurgaon 122002, Haryana

Distributed in Canada by Publishers Group Canada,
76 Stafford Street, Unit 300
Toronto, Ontario M6J 2S1

ISBN: 978-178578-832-1

Typeset in Iowan by Marie Doherty

Printed and bound in Great Britain
by Clays Ltd, Elcograf S.p.A.

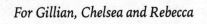

For Gillian, Chelsea and Rebecca

ABOUT THE AUTHOR

Brian Clegg is the author of many books, including most recently *Ten Days in Physics That Shook the World* and *What Do You Think You Are?* as well as *Quantum Computing* in the Hot Science series. His *Dice World* and *A Brief History of Infinity* were both longlisted for the Royal Society Prize for Science Books. Brian has written for numerous publications including *The Times*, *Nature*, the *Observer*, the *Wall Street Journal*, BBC *Science Focus* and *Physics World*. He is the editor of popular science.co.uk and blogs at brianclegg.blogspot.com

www.brianclegg.net

CONTENTS

CONTENTS

ACKNOWLEDGEMENTS

Thanks to the staff at Icon Books, notably Duncan Heath and Robert Sharman. Many years ago, I took an MA in Operational Research at the University of Lancaster, which introduced me to some of the concepts of game theory. My thanks to the lecturers there, particularly Graham Rand, who is still involved with the university. He edits the operational research magazine *Impact*, for which I have written many articles, including one that led me to look more into auctions and game theory.

GAMES AND
THE REAL WORLD

1

When I first bought a textbook on game theory many years ago, never having come across the term before, I felt cheated. I was expecting something fun that would tell me the optimal strategies for winning at card games, backgammon and Monopoly. I wanted an interesting analysis of how the games worked mathematically under the hood. Ideally there would also be guidance on how to create your own interesting board games. Instead, I found descriptions of a series of 'games' that no one had ever played, with tables of outcomes that did not so much give guidance as show just how impossible it often was to come up with a useful outcome. This was interspersed with a hefty load of mathematical equations. And yet, the more I read about game theory, the closer it seemed to one of my favourite classics of science fiction.

For his 1950s *Foundation* series of books (made into a TV show in 2021), Isaac Asimov came up with the concept of 'psychohistory'. This is an imaginary mathematical mechanism for predicting the future, based on an understanding of human psychology and the behaviour of masses of people. In practice, psychohistory was never going to happen. The repeated failures of pollsters who amass vast amounts of data to predict the outcomes of elections, or decisions such as the UK's Brexit referendum, make it clear that people form far too complex a system to enable reliable mathematical predictions of outcomes. Yet game theory does achieve some of the promise of psychohistory by resorting to the classic approach used by science, particularly physics: modelling.

The mathematical models used in physics reduce complex systems to simpler combinations of objects and their interactions. Messy aspects of the system are often ignored (it will be noted that this is happening). So, for example, Newton's familiar laws of motion at first glance don't appear to describe the real world very well. The first law states that an object in motion will keep moving unless acted on by a force. In everyday experience, such countering forces – like friction and air resistance – are ubiquitous; yet for convenience, models often ignore such things, as they add complexity and can be difficult to account for. This means that the model does not reflect reality – without friction and air resistance, once you gave it a push, a ball on a flat surface would roll on for ever. But simplification makes calculations more manageable and gives an approximation to reality. Similarly, game theory uses mathematical models that simplify human interactions and decisions as much as is possible to help understand those processes.

The theory of games started with the development of the mathematical field of probability to deal with gambling games and other pastimes where the outcome was dependent on a random source, such as the throw of dice or the toss of a coin. However, in the first half of the twentieth century, a handful of individuals and a quasi-governmental American institution took some of the basic mathematics of games and began to apply it to decision-making problems, ranging from economics to the best strategy to win a nuclear war.

The field that was developed under the name of game theory became detached from 'real' games. It was all about strategy – what was the best approach to win, given a set of choices available to two or more players. Games were transformed from pastimes to something deadly serious. This shift was so strong that often those who deal with game theory totally ignore what the rest of the world calls games. However, I believe that this is a mistake. Real games still form part of the continuum – it is just that many familiar games are not interesting from a game theory perspective, either because they are too dependent on random chance, with no strategy, or because they are too complex for strategies to be developed.

It's worth spending a moment on the 'strategy' word here, as it is often misused, and game theory has its own specialist meaning for the term. A strategy is a plan to achieve a goal. However, as J.D. Williams pointed out in his light-hearted 1960s book *The Compleat Strategyst*, in game theory, a strategy 'designates any complete plan'. In general usage, a strategy is usually a best effort to achieve something. But in game theory, a strategy is *any* complete plan for playing the game, no matter how good or bad. In chess, for example, your strategy could be to always play the piece closest to the bottom left-hand corner of the board that is available to move. Such a strategy would pretty much guarantee losing, but it would nevertheless be a strategy in game theory terms.

Much early game theory was developed to deal with situations where two players went head-to-head in an aggressive win-or-lose situation. This was the circumstance, for example, facing American military strategists when applying game theory to nuclear warfare and whether it was better to be reactive or pre-emptive when it came to nuclear strikes (arguably more a lose-lose scenario than win-or-lose). However, the most valuable impact of game theory in recent years has been in the design of specialist mechanisms to deal with spectrum auctions.

Selling a spectrum

The word 'spectrum' suggests that these auctions are something to do with selling off an array of colours, but here a different part of the electromagnetic spectrum is under consideration: not visible light, but the segment of radio frequencies available for, usually, mobile phones and wi-fi.

Historically, radio bandwidth was primarily used for broadcast radio and TV, with a relatively small number of transmitters sending signals to many receivers. Because of overlaps between different transmitters and applications, and because of the crude technology originally used, wide swathes of the radio bandwidth were allocated to broadcasters.

The exact definition of radio is a loose one. The electromagnetic spectrum is divided up by frequency or wavelength. Wavelength is the distance between equivalent points in the repeating cycles along the progress of a wave. Frequency is the number of such complete cycles of the wave that take place in a second.

Frequencies on the entire electromagnetic spectrum – which includes radio, microwaves, infrared, visible light, ultraviolet, X-rays and gamma rays – vary from a handful of

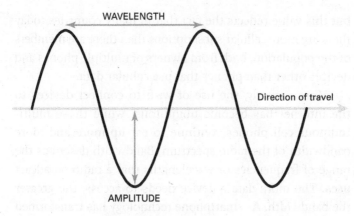

Figure 1.1. Structure of a wave.

hertz (cycles per second) through to hundreds of exahertz, where an exahertz is a million trillion hertz. The equivalent wavelengths run from hundreds of thousands of kilometres to picometres (trillionths of a metre).

Radio comes at the bottom end of the spectrum, with the lowest frequencies and longest wavelengths, at its highest reaching wavelengths of about 1cm and frequencies of hundreds of gigahertz (a gigahertz is a billion hertz), though signals at the top end of the radio range are often referred to as microwaves, first employed for communications and radar, but now also used in the eponymous ovens.

What has transformed the need to squeeze every bit out of the radio spectrum is the growth of two applications – cellular phones and wireless internet. Worldwide cell phone ownership has risen dramatically. In the mid-1990s, around 5 per cent of the world's population had access to a cell phone. By 2015, the 100 per cent mark had been passed. It might seem that only sportspeople and competitors in TV game shows claim to be able to give more than 100 per cent, but this value reflects the fact that in many countries today there are more cellular subscriptions than there are members of the population, both from owners of multiple phones and devices other than phones that use cellular data.

More recently, the use of wi-fi to connect devices to the internet has become ubiquitous, while those multitudinous cell phones continue to eat up more and more bandwidth of the radio spectrum. Bandwidth describes the range of frequencies or wavelengths that a radio broadcast uses. The more data a device needs to access, the greater the bandwidth. As smartphone technology has transformed cell phones from being simple communication devices to powerful pocket computers, they are starting to use the high-bandwidth flows of data needed to stream videos and perform other data-intensive tasks. This requirement has seen a rapid transition through 3G (third generation) and 4G connections, with 5G now becoming available, providing data rates that had previously only been possible through fixed fibre-optic connections.

At the same time, television, one of our biggest historical consumers of radio bandwidth, is undergoing a two-part revolution. The first change was from analogue to digital. Digital channels take up a lot less bandwidth than their analogue equivalents, because the data is compressed before transmitting it, making it possible to free up more frequencies for mobile data access. The other stage of television's transformation, which is only just starting to have a major effect but will transform TV viewing forever, is the move from broadcasting to streaming. Already, a percentage of the population watch most of their TV over the internet. In time, all TV will be watched this way and the bandwidth occupied by TV will be released.

Monetising bandwidth

An example of the process of transferring parts of the TV spectrum to mobile usage in America gives a dramatic portrayal of the way that game theory has come to play a major role in what can be a very lucrative process for governments.

In 2017, the US Federal Communications Commission (FCC), which regulates US telecommunications, realised there was an opportunity to reshuffle many TV stations' frequencies, freeing up bandwidth for mobile data. Specifically, they looked at the top end of the 600 MHz TV band, traditionally known as UHF (ultra-high frequency). This proved a particularly useful segment of bandwidth as it was adjacent in the spectrum to existing mobile phone bands, has good range and is effective at penetrating the walls of buildings, which is something of an essential for mobile signals.

The technical teams responsible for making this happen had two challenges: ensuring that the requirements for TV signals were still covered, though potentially on different frequencies; and getting the most money from the telecoms providers who wanted licences to use more of the available bandwidth for their customers.

The optimisation of the TV channel allocation made use of a sophisticated mathematical algorithm, but from the game theory viewpoint, the interesting part of the process was the mechanism for allocating licences to the mobile phone operators. The FCC would use an ancient mechanism for selling items among multiple competing interested parties, the auction – but with a new twist devised using game theory.

Remember that game theory is about more than playing traditional games – it's a mechanism for designing strategies and for decision-making when taking on opponents. Taking part in an auction is exactly the kind of process that game theory was designed to handle: bidders are competitive 'players' in a game where the prizes are (in this case) access to bandwidth. How effective a strategy can be often depends on how much we know about the desires and strategies of our opponents. The degree of information available is crucial to the way the game plays out, and this has become central to the design of sophisticated auctions. Before seeing how this is done, it will be helpful to take a look at an apparently simple game that influenced the development of game theory – poker.

Information and games

If, like me, you aren't a poker player, you may be surprised at the suggestion that poker is simple, because it can be tricky to remember the priority of the different hands. However, given those rules, the play is very straightforward – a hand with a higher value always wins.

Unlike most card games, poker has many different formats. In some, known as 'draw poker', the players' cards are concealed. The only source of information a player has about the strength of the hands of his or her opponents is the way that the players bet and anything that can be deduced from their speech and body language. However, other formats, such as stud poker (where some of a player's cards are dealt face up) and Texas hold 'em (where cards displayed on the table are included in every player's hand), provide players with some information on what is available to their opponents.

Imagine, though, that all cards in every hand were always visible. It wouldn't be much of a game, as everyone would have perfect information on what the other players were holding and therefore would know exactly what they were likely to do (unless they were very silly). The level of information available has a strong influence on the ability of the players to develop appropriate strategies.

We tend to think of an auction as a marketplace, but its power is as a mechanism for sharing information. It exposes the preferences of players of the auction game (the bidders), showing how far they are inclined to go to gain a particular outcome. Unless they get carried away and become irrational, players of the game will not bid higher than they consider the item on sale to be worth. And this is crucial information, because initially no one knows what an object is worth. We are used to many things having a list price, but this is an arbitrary convention. In reality, something on sale is worth what someone is prepared to pay for it. With a list price, the vendor can only guess what that amount is and see if anyone will buy. But an auction is a vehicle for establishing what that value is among the community of game players.

In the 2017 US spectrum auction process, the gaming power of the auction would be used doubly, first on the TV companies and then on the mobile phone networks. The first phase was to use auctions to see what value TV companies put on releasing bandwidth. The companies were paid to free up some of their frequencies and move to new ones. This involved a kind of auction known as a reverse auction, where, unlike a conventional auction, there is a single buyer and multiple sellers. Each TV station was offered a tailored starting price. Robert Lees, director of the Smith Institute at Harwell, Oxfordshire, who was involved in the FCC process, noted: 'Initial levels were set taking into account the coverage areas and coverage populations of each TV station. So, stations covering large areas of urban population would see high initial prices in the reverse auction. Another consideration was to set the prices at levels which would be sufficiently attractive to encourage high participation levels from the broadcasters.'

If a station accepted the offer, they stayed in the auction process. If they rejected it, they dropped out and could keep their existing bandwidth but had to move, without compensation, to a new channel (which would cost them money to undertake as they would need to change broadcasting equipment and retune customers' TVs). In the next round, the prices were lowered and the process repeated. Eventually there would be no new channels left to move to. At this point, the auction stopped and the remaining stations were paid the offer at this level to give up their bandwidth, provided there was sufficient funding available to cover the offers.

When the required amount of bandwidth had been freed up, the second style of auction, a more conventional 'forward' auction was started between the FCC and the mobile networks. Sections of the freed-up bandwidth were given a starting price, and anyone prepared to pay the requested amount entered the auction. The amount then went up, with players dropping out as the pricing got too rich, until the bandwidth was allocated to the last player standing.

This wasn't quite the end of the process, as it was possible that bidding would have dropped off too early to be able to fund the release of the TV channels. If this were the case, the auction would be abandoned and restarted with smaller units of bandwidth until a successful outcome was reached.

There had been plenty of spectrum auctions before, but what was unique here was the two-way facing auctions. Mobile phone companies were familiar with the auction process, but it was new to the TV companies.

According to Robert Leese, 'The FCC spent a great deal of time with [the TV companies] to make sure that they understood the process and that all their concerns were addressed. One key feature of the auction design was that participation for the broadcasters should be as easy as possible. They were never asked to select from more than three options at a time. Another key feature was that broadcasters were free to drop out of the process at any time (or not to participate in the first place), safe in the knowledge that they would not end up in a materially worse situation with regard to interference than their situation before the auction.'

Game theory was crucial to devising this auction design, as it is surprisingly easy to get auctions wrong. As we will discover in Chapter 6, by not anticipating the strategies of bidders, some spectrum auctions have been disasters. However, in the case of the 2017 FCC process, the auctions raised over \$10 billion to be paid to the TV industry for the released bandwidth, as well as over \$7 billion surplus for the government.

We will see how game theory got to this point in history – and how it has moved on thanks to the ability of computers to repeatedly play games – but to get into the basics of game theory, we need to travel back in time to the first mathematical explanations of games of chance.

PLACE YOUR BETS   2

The original theories of games were based on the mathematics of chance – probability. Some of the games studied were purely probabilistic, others combined chance with strategies and decision-making.

The simplest probabilistic games are those using the toss of a coin. Coin-tossing has the benefit of requiring a minimal level of game equipment, while providing a rather beautiful mechanism in the spin of a coin in the air. Strictly speaking, a coin toss is not entirely fair as it will typically result in the side facing up at the start of the toss being slightly more likely to be the outcome than the other side – but to a reasonable approximation, a fair coin generates a 50:50 probability of turning up heads or tails on any particularly throw.

The most trivial game based on a coin toss is simply to predict heads or tails. As this is purely random, there can be no strategy and as such it doesn't come under the remit of game theory. However, things get a lot more interesting when multiple tosses are involved. To understand what's happening, we need to follow the ideas of the Italian physician and gambler Girolamo Cardano, who wrote a book called *Liber de Ludo Aleae* ('Book of Games of Chance') – which was the first systematic exploration of probability in games.*

One of the innovations that would arise from Cardano's work was to represent chance using fractions. If we toss a fair coin, half of the time it will come up heads and half of the time it will come up tails. As a result, we can represent the probability of the outcome as ½ for heads and ½ for tails. The total probability of all options should always add up to 1. This numerical representation is extremely useful because it makes it easy to move from probabilities attached to a single event to dealing with multiple events. It's easier to briefly move away from coins to see how the arithmetic of chance developed.

Consider throwing dice. Each standard die has six possible outcomes: with a fair die each is supposed to be equally likely.** Each number from one to six has a ⅙ chance of turning up on a single throw. Cardano showed that to get the chance of any one of multiple outcomes, the probabilities are added. So, for example, the chance of getting either a one *or* a two with a single die is ⅙ + ⅙ = ⅓. The chance of getting, say, one, two or three is ⅙ + ⅙ + ⅙ = ½. Similarly, Cardano worked out that the chance of getting a six on a first throw, then getting a six again on a second throw (or with the simultaneous throw of two dice) was ⅙ × ⅙, or 1/36.

---

* Cardano's book was written in the 1560s, but would not be published until 1663, 99 years after his death, probably because of the then-prevalent view of gambling as disreputable.

** Just as it's difficult to be sure that a coin toss is fair, dice are often slightly biased towards the six. This is because the 'pips' indicating the number are often indentations, meaning the six side is slightly lighter than the opposite side with just one pip.

Cardano also took on the trickier concept of the chance of getting a six with *either* of two dice, or with either of two throws of a single die. Here, we can't just add the probabilities (otherwise, with six dice or six throws you would be guaranteed to get a six). We know that the chance of getting a six with one die is $\frac{1}{6}$, so the chance of not getting a six is $\frac{5}{6}$. This means that the chance of not getting a six with one die and then not getting a six again on the second throw is $\frac{5}{6} \times \frac{5}{6}$, making it $\frac{25}{36}$. If that's the chance of not getting a six at all, then the chance of getting at least one six with two throws is $1 - \frac{25}{36}$, which is $\frac{11}{36}$. It's just less than a $\frac{1}{3}$ chance.

This arithmetic of chance also allows us to start setting a strategy based on the outcomes produced by throwing two or more dice.* Once there is more than one die in play, different outcomes have different probabilities. With two dice, the most likely outcome is seven, which has a $\frac{6}{36}$ (or $\frac{1}{6}$) chance of turning up. By comparison, a two or a twelve only has a $\frac{1}{36}$ chance of being thrown, while a five or a nine has a $\frac{4}{36}$ ($\frac{1}{9}$) chance. Knowledge of these probabilities has a strategic role in games that involve throwing two dice, such as backgammon or Monopoly.

The hidden strategy

Returning to coins, the importance of having the right strategy becomes more obvious with a more sophisticated game than a single toss. Let's imagine that the players' goal is not to make a particular throw – heads (H) or tails (T) – but to

---

* Although relatively uncommon now, many early dice games depended on the outcome of throwing three dice.

get a particular sequence of heads and tails. Imagine, for example, that players are allowed to choose a sequence of three outcomes and a coin is then repeatedly tossed until that sequence crops up. The number of tosses required to get to this sequence gives the player their score – after everyone has had a go, the player with the lowest score wins.

Given the fact that heads and tails turn up with equal chances, you might imagine that it is no more possible to devise a useful strategy here than it is with a single coin toss. But let's consider what happens if a person is trying to decide between the sequence HTT and the sequence HTH. If a coin were just tossed three times, each of these sequences has exactly the same chance of coming up – ⅛. It doesn't matter which the player chooses. But that's not the game. The rule here is that we continue tossing until the chosen sequence emerges. And in this case, an appropriate strategy gives a player an edge. Given those options, choosing HTT is better than HTH. This emerges from thinking through what would happen if things went wrong.

In either case, you don't have a chance of winning until HT is thrown, whereupon, if your choice comes up next, you have won – but the outcome is different if the third throw doesn't go your way. Let's imagine that you choose HTH, but the first three throws are HTT. To get as far as HT again, you would have to throw an H and then a T – there's a ¼ chance of that happening. But if your strategy were HTT and the first three throws were HTH, then you already have H, the initial throw of the sequence, so now you only need to throw a T – with a ½ chance – to get back to HT. Counterintuitively, you are more likely to win by trying for HTT than you are for HTH. More generally, finishing the sequence with a different face to the one you start with is beneficial.

Games involving repeated tosses of a coin often require careful assessment of what the best strategy is, as demonstrated by the mind-bending nature of a game devised by two leading eighteenth-century mathematicians, the cousins Daniel and Nicolaus Bernoulli. The strategy for playing this game, as for many others, depends on a concept invented by Daniel Bernoulli, called expected value.

What's it worth?

Imagine you are offered two options for taking part in a coin-tossing game. You can either win £100 if you toss a head, or £200 if you toss two heads in a row. Which outcome is better? (In this unusually generous game, you get nothing for any other outcomes, but you don't lose anything either.) The expected value – also known as the expected return – is discovered by multiplying the outcome by the chance of getting that outcome. In the single-coin-toss version of the game, you get £100 with a ½ probability – so the expected value is £100 × ½ = £50. For the game requiring two tosses, you get £200 with a ¼ probability, making the expected value the same as the previous game, as £200 × ¼ = £50.

All things being equal, Bernoulli's concept of expected value means that you shouldn't care which game you choose – each has the same expected value. If you played the game many times over, you would expect to get roughly the same amount of money from either game. There's a devil in the detail, though. Playing the game once, you are twice as likely to win with the £100 version. Although the expected value of the two variants is the same, the strategy has to be influenced by another important concept devised by Daniel Bernoulli – the utility of the outcome.

The utility reflects how significant the potential gain or loss is to you as an individual. An amount of £100 would be considered very differently if you were a millionaire compared to if you were on the poverty line. If it's a not big deal to you whether or not you win some money in this game, you may well choose the £200 game, taking the extra risk for a bigger potential reward. But if it's more important to win something – anything – than it is to win big, you are better off going for the £100 game.

With these concepts in place, we are now ready to take on the Bernoullis' mind-boggling game. Here, you repeatedly toss a coin until you get a head, at which point the game finishes. If the first throw is a head, you win £1. If the second throw comes up heads, it's double the prize money: £2. If there's no head until the third throw, the money is doubled again: £4. If it takes four throws to get to a head you win £8 … and so on, for however many throws it takes. But unlike the pure generosity of the previous game, this game has an entry cost. The strategy required is to decide what you would be prepared to pay to play this game.

If the entry cost were 50p, the strategy would be trivial. You are bound to win at least £1, so you should definitely play. Even if the cost were £1, you might as well take part because you will get your stake back, whatever the outcome – you can't lose. But should you go higher than £1, and if so, by how much? We need Bernoulli's concepts of expected value and utility to work out your best strategy.

To calculate the expected value, you need to consider all the possible outcomes of the game, as it doesn't have a fixed length. There is a ½ chance of getting £1, so the first throw contributes 50p to the expected value. There's a ¼ chance of getting £2, so the second throw contributes an additional 50p to the expected value. There's a ⅛ chance of getting £4 – so it's 50p again. The total expected value is 50p for each of the infinite set of possible throws, making the total expected value infinite.* In terms of expected value alone, *whatever* the entry cost is for the game, it is worth playing.

When we bring in utility, though, things look different. The most likely win is £1. There is only a ¹⁄₁₂₈ chance of winning £128 or more. The higher the reward, the less likely it is. Clearly, no one but an impulsive billionaire would pay, say, £1 million to enter a game where the most likely outcome is to win £1. The chances of winning more than a million pounds are ¹⁄₁,₀₄₈,₅₇₆ – worse than a one-in-a-million chance. The strategy selected has to take into account the utility for the specific player. Depending on your personal worth, what you consider a trivial amount may be £1 or £1 million – but it would be a poor strategy to risk paying more than you can easily afford to lose on a game like this.

Despite the many schemes and systems that have been devised historically, there are no strategies available to those playing fair games of chance where the players can make no decisions, but simply wait for the outcome of a single coin toss or roll of a die. However, many other games exist where there is the potential for applying game theory.

In increasing complexity, we will take a look at noughts and crosses (tic-tac-toe), backgammon, Monopoly and Go.

---

* In practice a real game could not go on for ever, but we can say the expected value has no maximum size.

Losing should not be an option

Noughts and crosses demonstrates that requiring strategy does not imply complexity: this is a game where strategy is the only contributor to the outcome – there is no chance involved. In some two-player games, following the best strategy means that the first or second player can always win, but here the perfect strategy will always result in a draw. And that strategy is so straightforward that with a little experience, almost all players reach perfection. In case you somehow missed out on noughts and crosses when growing up, the game is very simple, played on a three-by-three 'board' (usually simply lines drawn on a piece of paper):

Figure 2.1. Noughts and crosses board.

Players take turns to place or draw an O or an X in one of the nine available spaces. If a player completes a line of three (horizontal, vertical or diagonal), they win. A good player aims to set up a position whereby they will be able to complete either one of two lines, so that their opponent can only block one of those plays. If both players adopt the best strategy, this is not possible and they will always draw.

The first move made by the second player can make the difference between a draw and a loss for that player. Wherever the first player goes, as long as the second player starts in a corner or the centre, it is always possible to force a draw. But if the second player starts in the middle of an edge, they can be forced to lose.

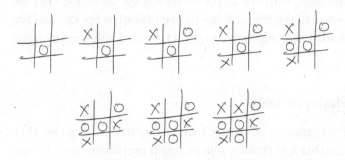

**Figure 2.2. A correct strategy from
both players, forcing a draw.**

In the example above, O starts and chooses the centre. X makes a correct response in a corner, and there follows a sequence of O setting up the first two of a row and X blocking until it is no longer possible to complete a row of three.

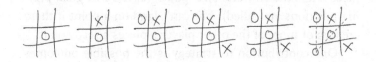

Figure 2.3. X adopts a losing strategy.

However, if X, playing second, chooses the middle of an edge, then O can take the dominant combination of centre and corner. Now X is forced to stop O's diagonal and O can add a third mark which now gives O two possible lines of attack – whichever line X blocks, O can win by completing the other.

What lies behind the failing strategy? If, in that second

game, X had placed a mark in the corner (as in the first game), that player would have two possible future directions to set up two in a row. However, by going in the middle of the edge, with one direction already cut off by the O in the centre of the board, X has halved his or her options and has made losing inevitable unless O makes a silly second move.

Playing the tables

Backgammon is a much more sophisticated game than noughts and crosses, one in which probability and strategy each have a part to play. This is an ancient game, with variants dating back thousands of years, and was historically known as 'tables'. The aim of the game is to move pieces around and finally off the board, based on the throw of two dice (the scores on which are counted separately, so throwing a six and a five, say, enables moves of six and five, rather than a single move of eleven). Players can knock off an opponent's piece if it is the only one on a 'point' (as the triangular playing positions are called), but can't land on a point if there are two or more of the other player's pieces already on it.

One contribution to strategy is the possible outcomes from throwing two dice. As mentioned above, seven is the most likely aggregate score (because it can be made up of 1+6, 2+5, 3+4, 4+3, 5+2 and 6+1); the probabilities of the other aggregates are listed in the table below:

2	3	4	5	6	7	8	9	10	11	12
$\frac{1}{36}$	$\frac{2}{36}$	$\frac{3}{36}$	$\frac{4}{36}$	$\frac{5}{36}$	$\frac{6}{36}$	$\frac{5}{36}$	$\frac{4}{36}$	$\frac{3}{36}$	$\frac{2}{36}$	$\frac{1}{36}$

Table 2.1. Probabilities of aggregate scores using two dice.

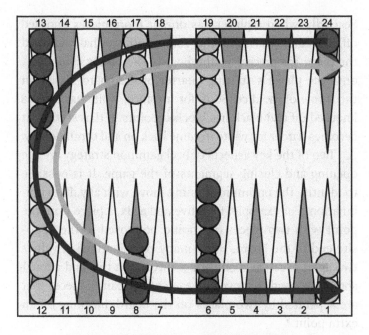

**Figure 2.4. Backgammon starting position,
showing direction of movement.**

Knowing this is useful, as one of the gambits available is
to apply the scores of both dice to the same piece – so,
for example, you might move a piece six places, and then
another five places, achieving the aggregate of eleven.
However, important though this probability is to the outcome of the game, its significance is modified by the role
played by blocked-off points. Much of the strategy of the
game involves the manipulation of these blocks. This is particularly important for two reasons. Firstly, because the value
on each die has to be moved individually, a piece can only
be moved eleven (sticking with the example above) if either
the point five ahead or the point six ahead is not blocked.

Secondly, when a player has one or more pieces knocked off, they cannot take any other action until they have returned the piece (or pieces) to the board. The pieces come back onto the player's starting quarter of the board, based on the throw of the dice. So if, for instance, white has a piece knocked off and black has blocked points in that section, it becomes harder for white to come back on and continue play.

Two of the key aspects of backgammon strategy are the opening and closing segments of the game. It is possible to identify the optimum opening move with any dice combination. For example, for a five and a six, a piece from the point with two pieces on should be moved to the opposite end of the board, while many combinations of values two apart (such as a one and a three, or a two and a four) should be used to move one each of the pair of pieces that are separated by two places on the board, blocking off an extra point.*

Similarly, when finishing the game, a player has to get all their pieces into their final quarter of the board before they can begin moving them off. Often there is a choice between moving two pieces a short distance or one piece a long distance. If the two-piece move enables the player to get two pieces onto the final quarter, or two pieces off the board entirely, that is strategically preferable, as it leaves open the possibility of finishing in fewer moves.

---

* Some of the optimum opening moves in backgammon are disputed. For example, whereas most throws two values apart will definitely be treated as described by a good player, some of those players prefer alternatives for the combination of six and four.

Simple though it looks, there are about 100 million billion possible positions available on the backgammon board, and the strategic range is increased by the most sophisticated addition to the gameplay: the ability to 'double'. By default, winning a game gains the player one point (or unit of currency). This value is increased to two points if a player finishes before the other has got any pieces off the board (known as a gammon) or three points if a player finishes while the other still has at least one piece left in their first quarter of the board (called a backgammon).

However, there is also the opportunity for either player, at the start of their turn, to double the value of the game. If the other player accepts, the points available to the winner double – if the other player doesn't accept, the player offering the double immediately wins, receiving the current value of the game. Once one player has doubled and has been accepted, only the other player can double – this ability passes back and forth between the players.

Doubling is conventionally recorded on a 'doubling die' which has the numbers 2, 4, 8, 16, 32 and 64 on its sides – however, there is no limit in the rules to the doubling, which combined with the fact it is possible to repeatedly knock pieces off, resetting their position, means that in principle there is an infinite set of possible states of the game. (The calculation for the number of possible positions above assumes that there are only three possible states of doubling, corresponding to no one having doubled, white having control of doubling or black having control of doubling).

Compared with noughts and crosses, then, although backgammon is not purely strategic, it does offer many opportunities for making use of mathematical concepts to enhance strategy.

Advance to go

Many modern board games start from a structure inspired by the backgammon board, though often each player only has a single piece and the board is treated as a continuous loop that can be repeatedly traversed. But in most such games, some or all of the positions on the board have distinctive properties. This is literally the case with probably the best-known twentieth-century board game, Monopoly. Although a version was first devised in 1903 (to demonstrate the evils of property ownership), the game was commercially produced in its familiar form in 1935, based on the streets of Atlantic City, New Jersey. A London version came out in 1936, followed by many other locations around the world.

As with backgammon, the use of two dice makes the varying probabilities of different combinations coming up a factor in the strategy of the game, though, as we will see, in Monopoly these probabilities are best employed by working backwards from the locations of specific squares.

Monopoly players may purchase squares on the board that they land on, subsequently charging other players for landing on those squares – a charge that can be increased when a player owns a matching set of squares, particularly if he or she invests by building properties on them. A good move when choosing where to build is to consider which properties the other players are more likely to land on. As we saw on page 22, the most likely throw with two dice is a seven, while anything from five to nine has a relatively high probability of occurring.

The table above for the chances of different throws is still approximately correct, but the distribution with Monopoly is skewed because when a double is thrown, the player throws again. In principle this can happen twice in a row (if it happens three times, they go to jail), so there are significantly more places a player may be forced to land during their next turn. This also alters the probabilities, meaning that, say, the square eight places ahead is more likely to be landed on than the square six places ahead, whereas with a single throw of two dice there is an equal chance of landing on these squares. This is because there are more ways of getting eight from two throws of two dice than there are of getting six. But the key throws of five to nine still dominate outcomes.

Unlike backgammon, in Monopoly there are other ways to get a piece to move around the board than using the dice. One is the 'luck' cards Chance and Community Chest. Each of these has a range of outcomes, which may involve gaining or losing money – but in terms of strategy, the important factor is that they have the potential to move a player to a specific square. This makes those squares more valuable to own – specifically, squares such as the railway stations, Trafalgar Square/Illinois Avenue and Mayfair/Boardwalk.

Most players will at some point end up on the jail square (either visiting or actually in jail). This is because there are a range of ways to end up in jail, whether it's landing on the 'go to jail' square, getting the relevant Chance or Community Chest card, or throwing three doubles in a row – all on top of simply visiting by landing on the jail square in the usual way. This increases the probability of players landing on the squares that are between five and nine places ahead of jail once they leave it – so these are excellent properties to own and to build on. As a result, the station following jail plus the orange squares are more attractive than many other locations, while the magenta squares (other than the first one) and the red squares also get a boost.

There are other strategic subtleties that can be assessed by looking at the probability of the return on investment on building houses (building three houses at once is the most efficient approach). But what certainly is the case is that there is far more game theory benefit available in playing Monopoly than first appears to be the case.

Advanced Go

With some games of complex strategy, attempts have been made to codify that strategy so a computer can play; the earliest such board game was chess, but the pinnacle that proved harder to crack was the game of Go. This apparently simply game involves players taking turns to place either white or black stones on the crossing points of a rectangular grid, forcing the removal of the other player's pieces they completely surround, in a combinatorial explosion of possibilities.

In chess, for example, the white player has twenty initial moves to choose from (sixteen moves for pawns and four for knights); each of these 'openings' has been analysed at length. A standard Go board has 361 points a player can start on, and the alternatives within a few more moves spiral out of consideration. Go is estimated to have around 10^{170} possible positions, compared with chess, which is thought to have around 10^{50} possible positions.*

There are known strategies in the game that will immediately plot a course to defeat a novice player who places stones at random. These strategies often involve keeping

---

* 10^n is 1 followed by n zeros. So, 10^6 is a million. 10^{170} is 1 followed by 170 zeros, far more than the estimated number of atoms in the universe.

the player's own stones connected while attempting to cut off the opponent's stones – making the corners particularly attractive starting points. However, early attempts to produce high-level Go computer programs based on theory of play proved ineffective. Go's combination of a simple format and extremely complex combinations would require a different approach. When software was eventually developed that proved capable of beating champions – a program called AlphaGo – it was through a methodology that abandoned the development of conventional strategy.

AlphaGo makes use of a neural network, a computer structure that to a degree emulates part of the structure of the brain to provide a mechanism that can take decisions without being aware of what it is trying to do, let alone knowing the available strategies. In actual fact, the first version to beat a world champion combined training from human experts with self-learning – so strategies were involved. But in 2017, the team produced an updated version. To quote their article in *Nature*:

> Here we introduce an algorithm based solely on reinforcement learning, without human data, guidance or domain knowledge beyond game rules. AlphaGo becomes its own teacher: a neural network is trained to predict AlphaGo's own move selections and also the winner of AlphaGo's games [...] Starting tabula rasa, our new program AlphaGo Zero achieved superhuman performance, winning 100–0 against the previously published, champion-defeating AlphaGo.

Literally, all this version had to go on were the rules of the game, with no concept of strategies. By simply playing against itself nearly 5 million times, initially making moves at random, AlphaGo Zero transcended what had been possible previously. In reinforcement learning, the software is not told what to do – it is simply rewarded when things are going well, based on estimating the probability of winning from the current position, adding extra weighting to successful approaches it develops. The outcome can often be decisions that make no sense to Go experts – in fact, part of AlphaGo's strength is the way its plays can come out of the blue.

In a sense, AlphaGo does make use of strategies – but it uses no theory to develop them, nor is there any understanding of why a particular strategy is used. It's all down to trial-and-error learning (with emphasis on the 'error' initially). AlphaGo knows nothing about game theory.

The success of this approach identifies one of the limits of game theory. Go is simply not a game that can sensibly be addressed by the theory because it is far too rich in possible moves and counter-moves. It's rather similar to the way that the physics of a box of gas is handled. Although you could in principle work out the Newtonian behaviour of each gas molecule (leaving aside for the moment the oddities of quantum physics), in practice achieving this for the trillions upon trillions of molecules in any reasonable-sized box is beyond calculation and statistical approaches are used instead. But this does not in any way negate Newton's laws.

A different kind of game

There has been some speculation that the success of the AlphaGo software meant that artificial intelligence would now be able to expand to master practically any possible field. However, bear in mind that although Go is an intensely hard game to master, its rules are extremely simple. When mathematicians made the leap from board games to reality, a different approach was required, one that took in the more complex rules of the real world and achieved a modelling approach that kept the situations simple enough to deal with, even though the rules of human interaction are far more complex than any board game.

This focus would mean that the games that became central to game theory bore little resemblance to a board game or gambling game, becoming more like the kind of decision-making challenges that face human beings every day. In this sense, poker, which is sometimes described as the inspiration behind game theory, provides the connection between traditional games and game theory, because poker is one of the few games where bluffing is a major part of the player's armoury.

As John von Neumann points out in the book *Theory of Games and Economic Behavior*, bluffing has two purposes: to suggest you hold a weak hand when the hand is actually strong (so others will bet against you and lose); or to suggest you have a strong hand when it's weak (so others will fold and let you win). Applying game theory suggests that the best approach is to combine regular betting on strong hands with occasional betting on a weak hand.

In many card games, a command of probability is all you need to maximise your chance of winning. Famously, for example, blackjack (with variants also known as pontoon or 21), is a game where probability is key. In the game, the player(s) and the banker take cards from a shuffled pack, aiming to get as close as possible to 21 without overshooting (aces count as one or eleven to taste, while all face cards are ten). There is an element of basic strategy, which could include setting a minimum score to stick, rather than take another card. However, there is a central strategy to guide this decision by gaining extra information.

When cards have been played, they are discarded, not returned to the pack. This means that as the game progresses an observant player will have an increasing amount of information about the remaining cards still to come into play. For example, if playing with a single pack (usually in casinos multiple packs are used), once four aces have been played, there will be no more aces available. Awareness of which cards are removed from play makes it possible to calculate the changing chances of having various hands, giving guidance on when to stick with the cards you have.

This technique is known as counting cards. Bizarrely, although this approach depends on pure skill, with no attempt to mislead or cheat, casinos are allowed to treat it as breaking the rules. But the point is that probability allows a player with a good memory and command of the mathematics to decide what to do to maximise their chance of winning. However, in poker, if a player were to rely solely on probability to devise strategy, their bets would soon enable other players to get an idea of the strength of their hands. It is the ability to bluff and act as if you have a good hand when you don't (or vice versa) that makes all the difference. And it was this move from a purely probabilistic theory of games to one that incorporates behavioural strategies that made game theory new and interesting.

The man who was primarily responsible for this development is generally regarded as one of the most versatile applied mathematicians of the twentieth century. His name, which we have already encountered above, was John von Neumann.

VON NEUMANN'S GAMES

3

John von Neumann is probably less familiar than the English mathematician and codebreaker Alan Turing, but the two men were equally important in their contributions to the development of the computer, and von Neumann was also central to establishing game theory as a discipline. This seems to have come out of an interest (if no great skill) in playing poker; but, as we have already seen, it would develop far beyond the conventional concept of games.

We have so far taken it for granted that games can provide an analogy for decision-making in life. The Polish-British mathematician and intellectual commentator Jacob Bronowski, who worked with von Neumann during the Second World War, commented in his masterpiece *The Ascent of Man* that 'You must see that in a sense, all science, all human thought is a form of play. Abstract thought is the neoteny of the intellect,* by which man is able to continue

---

* Neoteny refers to the tendency to maintain juvenile characteristics or behaviours into adulthood. Humans are considered physically neotenous because they have a number of physical features (such

to carry out activities which have no immediate goal (other animals play only while young) in order to prepare himself for long-term strategies and plans.'

Born in Budapest, Hungary in 1903, as Neumann János, John von Neumann (later known as Johnny) took an interest in mathematics at an early age. This is more certain than frequently-made claims that, as a small child, von Neumann was capable of emulating his grandfather's party trick of performing effortless mental arithmetic on prodigiously large numbers: although in later life he did manage impressive arithmetical feats, it seemed to come after great mental exertion. However, he did have a remarkable memory, absorbing books at a rapid rate, with the ability to regurgitate their contents later. When the boy was ten, his father Miksa was awarded a hereditary title, resulting in the addition of 'von' to the German version of the family's surname.

In 1921, aged seventeen, von Neumann simultaneously set about studying for a degree in chemical engineering in Zurich (Miksa's preference for his son's career) and working on a PhD in mathematics through Budapest University. His choice of PhD topic was challenging – to untangle the mess that beset the axioms of set theory. Developed by German mathematician Georg Cantor, set theory is considered the foundation of arithmetic, but there was a problem with one of the axioms, the assumptions that are required for any mathematical proof to be constructed.

One of the set theory axioms, the so-called axiom of choice, read: 'For every set we can provide a mechanism for choosing one member of any non-empty subset of the set.'

---

as large heads and a flattened, hairless face) that other apes lose in adulthood. Here, Bronowski suggests that abstract thought is a continuation of the playfulness of young animals.

This seems straightforward, but there was no indication of how this choice was to be made. The way the axioms were set up led to potential confusion, but von Neumann managed to add an extra axiom, known as the axiom of foundation, that excluded the consideration of sets that would make set theory unmanageable. It didn't fix the problem that the axiom of choice remained independent from the other axioms, but it did make sets rigorously usable as Cantor had intended.

In the late 1920s, von Neumann would make mathematical contributions to the newly developed quantum mechanics while at the universities of Göttingen and Berlin. In 1929, he moved to Princeton University in America, soon afterwards marrying Marietta Kövesi, a childhood friend from his Hungarian home town, a marriage that would not last ten years. Von Neumann remained in America for the rest of his life, joining the Institute of Advanced Study in New Jersey (the American academic home of Albert Einstein) in 1933 and becoming an American citizen in 1937.

Von Neumann wrote his first, impressively influential paper on game theory in 1928, while still in Europe. Titled *Zur Theorie der Gesellschaftsspiele* (On the Theory of Games of Strategy), this established von Neumann's minimax theorem (see page 46), one of the key foundations of game theory.

His work on the topic culminated in his writing the intimidating 1944 tome *Theory of Games and Economic Behavior*, co-authored with economist Oskar Morgenstern. As we have seen, there had been a significant amount of work on the mathematical theory of games of chance going back to Cardano. A number of mathematicians, from the eighteenth-century British diplomat James Waldegrave onward, had made passing references to concepts such as game strategies over the years. By the 1920s, French mathematician Émil Borel showed the first signs of moving on to a version of game theory incorporating bluffing, and which had an application to political and military decision-making. But it was von Neumann who forged the accepted modern theory as a mathematical discipline.

Von Neumann appears to have been a driven individual whose work overshadowed his family life. Allegedly, when working on the *Theory of Games* book, which took up vast amounts of his time, his second wife Klara – also Hungarian-born, and who would go on to be one of the first computer programmers – commented, somewhat obscurely, that she did not want to have anything more to do with game theory (which was keeping her husband too busy) unless it included an elephant.

Von Neumann's sense of humour was often crude and sexist by modern standards, but he responded elegantly to his wife's challenge. In a section of the book on sets and partitions, sets are demonstrated with diagrams displaying a field of dots, which are broken up into sets and subsets by curved lines. One of these diagrams, only described in the text as demonstrating how lines are used to identify the elements of a partition (a part split off from the set that does not need to contiguous), contained the unmistakable shape of an elephant's head (see opposite).

Jacob Bronowski gives an insight into von Neumann's attitude to games, recounting a conversation they had during a taxi ride in London. When he heard the name of game theory, Bronowski, an enthusiastic chess player, assumed (as I later would) that von Neumann was dealing with games of that kind. 'No, no,' Bronowski tells us that von Neumann said, 'Chess is not a game. Chess is a well-defined form of

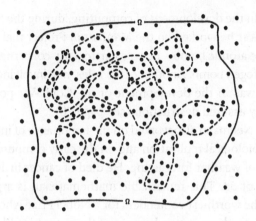

Figure 3.1. Figure from *Theory of Games*.
Reproduced courtesy of Princeton University Press

computation. You may not be able to work out the answers, but in theory there must be a solution, a right procedure in any position. Now real games are not like that. Real life is not like that. Real life consists of bluffing, of little tactics of deception, of asking yourself what is the other man going to think I mean to do. And that is what games are about in my theory.'

According to Bronowski, von Neumann's approach boiled down to making a clear distinction between tactics and strategies. Tactics are short term and are usually dependent on more detail than strategies. Strategies cover longer-term periods and can rarely be calculated exactly; but by using game theory, von Neumann realised it should be possible to identify better strategies, often selecting the best strategy for a given level of information.

Game theory forms only a small part of von Neumann's legacy. As well as being one of the two most important figures in the development of computing, during the Second World War he worked on the Manhattan Project that developed the atomic bomb – he was particularly concerned with the hydrogen bomb. Another of his roles, one involving game theory, was in the development of the Cold War policy of mutually assured destruction (see page 78).

Von Neumann also worked on the importance of information in biological replication and on the early computational science of weather forecasting. He died of cancer in 1957 at the age of 53. This remarkable mathematician is arguably one of the worthiest contenders for a Nobel Prize who never won one, perhaps in part because of the sheer breadth of his work. What lay behind much of it was bringing mathematical rigour – rationality – to a range of fields, including some, such as economics and decision-making, where numerical precision had not previously held much sway.

What is rational?

Game theory takes the logic and mathematical approach that makes it possible to produce the best strategy in games and applies the same approach to real-world situations. These can, in principle, range from everyday human interactions to key decisions in warfare. In practice, however, although many human activities can be treated as games, there are some significant limitations on game theory's applicability.

Broadly, these restrictions on the scope of game theory originate with two requirements: rationality and complexity. Game theory can only provide an implementable strategy if players act rationally. However, in the real world, exactly what is rational is far more complex than maximising one variable, which is the way that many simple games are constructed. An important lesson in this respect was provided by the UK's Brexit referendum to leave the European Union in 2016. In allowing a referendum, the strategists of the ruling Conservative party assumed that a rational decision would be the one that maximised immediate impact on the economy. There was little doubt that Brexit would make a financial hit on Britain's economy, so it was assumed that voters would act 'rationally' and vote to remain in the EU.

Unlike the economists and politicians, however, the voting public had a wide range of reasons for voting for Brexit: anything from regaining control of the country from an ever more bureaucratic European Union, through concerns about the perceived impact of uncontrolled immigration on their communities, to a desire to punish a metropolitan elite that, it was felt, treated provincials with disdain.

Even as this book is being written in 2021, practically every day we see ardent remain supporters missing the point and noting (almost celebrating) the impact on trade that has inevitably been felt after the completion of the withdrawal. Such people see those who disagree with them as irrational and stupid. But they miss the point that rationality is a multi-dimensional concept, and as such, those who wish to take on politics with game theory need to have a better understanding of the priorities of people outside their own sphere of experience.

Complexity and chaos

Brexit was a far more complex decision than many like to portray it, leading us to the second limitation of game theory.

The games at the heart of the theory often involve two players, each of whom has two possible options to choose between. There is no grey area, no room for negotiation. The rules are clear and precise. This is a perfect mathematicians' model world, but can bear little resemblance to reality. Games usually have tightly specified rules. But not only are there often many options and few clear rules in real-world situations, the environments in which decisions are made are also often mathematically chaotic.

Chaotic systems can appear random, even though they are purely deterministic, with clear chains of cause and effect. Unfortunately for those hoping to predict the behaviour of chaotic systems, very small differences in the starting position can rapidly produce very big changes in outcome. This is why weather forecasting has proved so difficult. Weather systems are chaotic – they are where the mathematics of chaos was first discovered. But the same chaotic nature applies to most systems of human interaction.

Of course, some very formalised interactions do have rules that are good enough for game theory to become accurate, but many are extremely difficult to pin down. However, this inherent complexity does not make game theory useless. It can be extremely valuable to get a better understanding of the dynamics in a decision process, seeing how one or more factors might influence the decision outcome if the players behave rationally based on those factors. We need to think of game theory as a *model* of a more complex system that delivers insights, not a means to reach a definitive correct answer. The model helps us explore influences, even if it does not always deliver an optimum strategy.

Why we need the maths

To mathematicians like John von Neumann, maths was the obvious tool to explore decision options and strategies. But to many of us, it isn't initially obvious why, when the idea of better understanding games is introduced, there is any need to go to the extent of using mathematics. There can be a suspicion that this is an attempt by mathematicians to ruin perfectly good pastimes by adding in a numerical element. Many people find maths an uncomfortable (or, dare I say it, boring) subject and can fail to see why common sense and the rules of the game are not enough to provide a winning strategy. After all, we surely try to be rational people. What could possibly go wrong?

A useful counter to this viewpoint is the public response to the game that is generally known as the Monty Hall problem. This challenge is based on an American gameshow from the 1960s called *Let's Make a Deal*. The final segment of the game, known as the Big Deal, involved players trading everything that they had won so far for the unknown prize behind one of three doors. Typically, one of these prizes was worth less than their accumulated winnings, one a little more and one a lot more.

In the modified version of the game which provides this problem, named after the host of *Let's Make a Deal*, Monty Hall, there is a twist. Here there is only one good door, while the other two are duds. The game is often presented as having one door with a sports car behind it, while each of the other two doors hides a goat. The player freely chooses one of the three doors (perhaps guided by the shouts of the audience). But before the prize is awarded, the host opens one of the other two doors, showing a goat. The player is now given the choice of sticking with their original choice, or of swapping to the other unopened door.

The strategy we are looking for here is to decide if the player is better off sticking with their original choice, is better changing to the other unopened door, or has no reason to prefer one door over the other. This puzzle was presented to the public in the 9 September 1990 edition of *Parade* magazine. For non-American readers, this is a Sunday newspaper supplement that accompanies more than 700 US publications, making it the most widely read magazine in the country, with an estimated readership of over 50 million. The problem turned up in a column called 'Ask Marilyn', written by Marilyn vos Savant, who was listed in the Guinness Book of Records as having the world's highest recorded IQ of 228. (She was in fact born Marilyn Mach; the surname she adopted professionally was her mother's maiden name.)

The Monty Hall problem was sent in to vos Savant by a Craig F. Whitaker of Columbia, Maryland. To most people, at first sight, the common-sense answer is that it doesn't matter which door you choose. After the host shows us a goat, there are two doors left to open, one with a goat behind it, the other with a car. That means the player has a 50:50 chance of winning, whichever door they choose. However, vos Savant answered that the first door chosen had a one-third chance of winning and the second door available had a two-thirds chance. The player should swap to the second door.

On 2 December, vos Savant returned to the problem, as she had been swamped with negative responses to her solution. As she put it: 'Good heavens! With so much learned opposition, I'll bet this one is going to keep math classes all over the country busy on Monday.' Comments included: 'Let me explain: if one door is shown to be a loser, that information changes the probability to ½. As a professional mathematician I am very concerned with the general public's lack of skills.' Another correspondent (like the previous one, a PhD) opined: 'You blew it, and you blew it big! ... There is enough mathematical illiteracy in this country, and we don't need the world's highest IQ propagating more. Shame!'

Vos Savant put forward further arguments as to why she was correct. These were more sexist times than the present and it would be interesting to know if a male columnist would have had as heated a response, but vos Savant was sent even more derogatory mail. Comments from academics, included: 'May I suggest that you obtain and refer to a standard textbook on probability before you attempt to answer a question of this type again?'; 'How many irate mathematicians are needed to get you to change your mind?'; 'Maybe women look at math problems differently than men'; 'You are the goat!'; and the delightful 'You're wrong, but look on the positive side. If all those Ph.D.s were wrong, the country would be in very serious trouble.' This last was from Everett Harman, PhD, of the US Army Research Institute.

In her final response on 17 February 1991, vos Savant noted: 'Gasp! If this controversy continues, even the *postman* won't be able to fit into the mailroom. I'm receiving thousands of letters, nearly all insisting I'm wrong, including one from the deputy director of the Center for Defense Information and another from a mathematical statistician from the National Institutes of Health! Of the letters from the general public, 92% are against my answer; and of the letters from universities, 65% are against my answer.'

But, as vos Savant went on to say, mathematical results aren't determined by votes. She reiterated her original argument and added a couple more. The first argument was to imagine there are a million doors, not just three. You pick door one. The host opens every other door but one – they're all goats. Wouldn't you switch to the remaining one? Another way of looking at it is that the car had a ⅔ chance of being behind one of the two doors the player did not choose. The host shows which of these other two doors not to open – he doesn't pick at random.

When I first heard about this problem, I was working in a roomful of applied mathematicians, who immediately dropped the jobs they were supposed to be working on and set about writing computer programs to simulate repeated plays of the game. They demonstrated without doubt that you were twice as likely to win if you switched. Yet so many people don't get it when the game is described. The problem defeats common sense: only a mathematical description can make it clear. And this is why game theory's maths is necessary.

Of course, the reader has to put the effort in. In her second column, vos Savant did lay out the kind of table that we will come back to repeatedly in exploring game theory, but most readers didn't understand it. Just working out what is happening in one of these tables can be a bit fiddly, but it is worth the effort because without following this kind of presentation, game theory will never make much sense.

Here's the game theory take on the Monty Hall problem (my version is a little more condensed than vos Savant's), assuming that the player initially selects door 1, which is how the problem was first posed. If the player switches, they will always switch to the door the host *didn't* open, as otherwise they are definitely switching to a goat (see Table 3.1).

The player wins two times out of three by switching.

Arrangement ↓ | Choice →	Switch	Stick
1: Car; 2: Goat; 3: Goat – Host opens 2 or 3	LOSE	WIN
1: Goat; 2: Car; 3: Goat – Host opens 3	WIN	LOSE
1: Goat; 2: Goat; 3: Car – Host opens 2	WIN	LOSE

Table 3.1. Outcomes for the player of switching and sticking in the Monty Hall problem.

Zero sum and win-win

There are two broad categories of game. In a zero-sum game – a term from game theory that has escaped into wider usage – in order for a player to win, another player has to lose. A game like noughts and crosses is zero sum. If I win, my opponent loses. Other games have the opportunity of win-win, where in principle everyone can win.* A simple win-win game we undertake each time we drive on the road is the 'Which side of the road should I drive?' game. Either left or right is a perfectly acceptable strategy where everyone wins, as long as every driver chooses the same option.

A more sophisticated example is a game that can be considered both zero sum and win-win. Take, for example, purchasing a loaf of bread. From a purely financial viewpoint, one person (the customer) loses money and one person (the baker) wins money. However, from the broader viewpoint of how the individuals involved in the game benefit (using the utility concept we first met on page 18), both have the potential to win – one gets a tasty loaf, the other some money.

---

* There is a third category of lose-lose games. It's hard to see why anyone would deliberately play these, though in real life, due to incomplete information or irrationality, such games are often played.

Most commercial transactions have this blended nature. This example shows how reality rapidly becomes more complex than the model provided by simple games, because making a purchase is a game with a *potential* to move beyond zero sum. It is only win-win if the price of the loaf is right for both parties. And for that matter, it's a requirement for a win-win that the loaf is edible and that the money is not forged.

Taking mini to the max

John von Neumann's best-known contribution to game theory (other than pretty much starting it as a serious mathematical field in the first place) is the minimax theorem. This only applies to two-person games that are zero sum, where the two players' desires for the outcome of the game are pulling in opposite directions. Von Neumann proved mathematically that in such games (unlike many that we will explore) there is always a best strategy for playing the game that is economically rational.

The theorem involves taking the worst possible scenario – the *minimum* opportunity for enrichment or reward – and then *maximising* the potential pay-off, producing the term *minimax*. Typically, that worst case is the result of your opponent guessing your strategy, so the opponent makes use of the strategy that makes you suffer most with each of your possible outcomes. This being the case, you can then choose the actual strategy that maximises your outcome (or at least minimises the pain).

A basic, but nonetheless elegant, example of minimax arises in the cake division problem. The minimax solution occurs when a dispute over how to divide a cake between two people is solved by asking one player to cut the cake and the other to choose how the resultant pieces are allocated. Assuming that the players are rational, the player doing the cutting will anticipate that the worst scenario that will inevitably play out is that the other player will take the larger piece. To maximise his or her own pay-off, therefore, the cutter must attempt to slice the cake into two equally sized portions.

Such two-player games with two possible strategies are sometimes referred to as 'toy' games, because few real-world problems are this simple. But the terminology is misleading. Some toy games are a valuable reflection of reality – cake-cutting being a good example. Others provide a useful model of a real decision – this does not invalidate the approach.

We can now put together a table of the choices available in cake division and the outcomes of making those choices, much as we did above with the Monty Hall problem. In the table below, player 1 is the person with the knife, player 2 the one who makes the choice. To encompass every possible game type we would have to show the outcome for both player 1 and player 2 in each box, but as this is a zero-sum game (there is only a fixed amount of cake), we can just show the outcomes for player 1, with player 2's outcome being the remainder.

The assumption I've made in naming the outcomes is that the result of cutting evenly is two pieces, with the one labelled 'Half+' slightly bigger than the one labelled 'Half–', due to the cutter's inability to make a perfect division. To explore the minimax solution, we add an extra column which features the minimum outcome of each row, and an extra row, featuring the maximum outcome of each column (shown in italics). This looks a little overwhelming, but once the approach is familiar, it becomes straightforward:

Player 1 ↓ | Player 2 →	Choose big	Choose small	Row minima
Cut evenly	Half-	Half+	**Half-**
Make one piece bigger	Small	Big	Small
Column maxima	**Half-**	Big	

**Table 3.2. Outcomes for player 1
in the cake-cutting game.**

Now comes the minimax bit, or more strictly the maximin/ minimax bit. Player 1 should choose the maximum of the row minima a.k.a. the maximin – in this case Half-. Player 2 should choose the minimum of the column maxima (the minimax) – here also Half-. These selections are highlighted in bold above.

As long as you remember to put minimum values at the end of each row and maximum values beneath each column, you won't go wrong. The reason, incidentally, that the minima and maxima are this way round is because the outcomes in the table are those for player 1 – they would be swapped if we gave the outcomes for player 2.

If the maximin and the minimax are the same, as is the case here, this is known as a saddle point and the intersection of the chosen row and the chosen column represents the best rational strategy for both players. In this particular case, the only sensible choice is that player 1 should cut the cake as evenly as possible and player 2 should choose the bigger of the two pieces (if they are distinguishable).

Mixing it up

To see how minimax works in a more typical game with numerical outcomes, let's take a simple matching problem.

In this game, each player has two items – say, a red ball and a blue ball. They each choose one of their balls. If they choose the same colour, player 1 wins, if each chooses a different colour, player 2 wins. The losing player gives the winning player £1. Unlike the cake-cutting game, here both players have the same strategies open to them (this is more common in simple two-player games). The resultant outcomes might look like this:

Player 1 ↓ | Player 2 →	Blue	Red	*Row minima*
Blue	£1	–£1	*–£1*
Red	–£1	£1	*–£1*
Column maxima	*£1*	*£1*	

**Table 3.3. Outcomes for player 1
in the colour-matching game.**

We appear to have a problem. There is no choice that produces a maximum of the row minima, or minimum of the column maxima. How, then, is it possible to produce a minimax strategy – something that von Neumann proved mathematically is available in *all* two-player, zero-sum games? We have to extend to the minimax of a 'mixed strategy', discovering what the result of playing multiple games would be. Just considering one game, the players can't devise a minimax strategy, but if the contestants had played a repeated game, an appropriate strategy emerges. Note that a mixed strategy is not about actually playing multiple games – more on what is involved with those in Chapter 5 – but about using what we can learn from hypothetical multiple games to advise how to behave in a single game.

In the worst case, if player 1 prefers blue, so chooses it

more often, and player 2 knows this, player 2 can get a net profit by always choosing red. Similarly, if player 1 is biased towards red, player 2 can win overall by always choosing blue. Accordingly, the maximin strategy for player 1 is to choose red and blue equally often, with a net outcome of 0 – no loss, no gain. This is the maximum minimum, as the minimum for choosing one colour is –1. The game is symmetrical, so assuming player 1 adopts the maximin strategy, player 2 should also choose red and blue equally often. More often, uncovering a mixed strategy will require more calculation, as we will see later.

But bear in mind that the game is only played once. This means that the mixed strategy has to be used probabilistically. A player can't choose a mix of blue and red (would that be purple?) on a single game. However, they can say that they should choose blue and red each with a 50 per cent probability – so the best approach is to toss a coin to choose a colour. In other games, it may be that the mixed strategy suggests one choice should be played more than the other – in this case the probabilities are adjusted accordingly, even though a single choice is going to be made.

To some observers, this is an uncomfortable aspect of minimax. Choosing at random, say as a result of tossing a coin, seems very unscientific. But to make use of another strategy is potentially worse because the opponent may have discovered what that strategy is to be. Only a random selection ensures that the game has its rational minimax outcome.

Note, incidentally, that minimax is not the best strategy in all circumstances – it's only the best option if your opponent is purely out for themselves, so acting in the way that's worst for you. In the above example, if player 2 always chose blue, then player 1's best strategy would not be the mixed strategy, but would be to always choose blue as well, so the colours matched, meaning player 1 would always win. As always with these games, the strategy is shaped by how much information each player has about the other's intentions. But if, for example, a player was inclined to choose blue because it was their favourite colour, they would be sensibly advised by the strategy to ignore this inclination and make a random pick, as it is always possible that the other player knows about their colour preference.

In a tree

To see what the outcome will be with a mixed strategy it can be useful to put together a tree of the decisions and apply probabilities to each part. In the matching game above, where both players adopt a mixed strategy of randomly selecting blue or red, the tree would be like this:

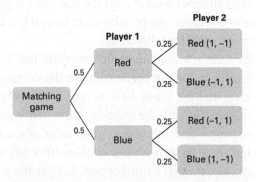

Figure 3.2. Tree for the matching game, with outcomes for player 1, player 2 in brackets.

To find the outcome, we multiply the probability of each out-come (which appears on the line leading up to the right-hand boxes) by the outcome if that choice is made by each player, then add them together. So, for player 1, whose outcomes are the first in the rightmost boxes (e.g. 1 where it says Red $(1, -1)$), we get $(0.25 \times 1) + (0.25 \times -1) + (0.25 \times -1) + (0.25 \times 1) = 0$. The equivalent calculation with the same outcome applies to player 2. This is trivial here – but it gets more interesting with more complex games.

It's a goal!

Whereas cake division is a deadly serious concept, the matching game above seems contrived and unlikely to occur in real life. And yet, slightly rewording the matching game turns it into an important factor in winning or losing many soccer matches – the penalty shootout. If we make player 1 the goalie and player 2 the penalty taker, then blue could be the ball going to the left and red could be the ball going to the right. The goalkeeper wins if both the ball and the goalie go in the same direction, the penalty taker wins if the ball and goalie move in different directions.

By the time it's obvious which way the ball is travel-ling, it is too late for the goalie to make the decision – the penalty taker and the goalie have to decide simultaneously. Traditionally, just as with the matching game above, the game theory solution – and the one adopted most of the time in real-life matches – is to randomly choose either left or right. But in this case, there is a third option that has the potential to benefit the penalty taker far more than the usual strategy – kicking the ball straight down the middle.

The reason this is such a good choice is that unless the goalie doesn't dive at all, kicking down the middle will always succeed, rather than winning half of the time.* However, it's a choice that real penalty takers hardly ever make. The reason is psychological. Kicking straight down the middle is an apparently stupid choice, because that's where the goalie is standing initially. If things go wrong and, for whatever reason, the goalie doesn't dive, the penalty taker will look stupid because he or she kicked the ball straight into the hands of the opposition. As a result, penalty takers don't make use of this option as much as they should. Let's have a look at the outcome table with three options:

Goalie ↓	Taker →	Left	Right	Middle	Row minima
Left		1	−1	−1	−1
Right		−1	1	−1	−1
Middle		−1	−1	1	−1
Column maxima		1	1	1	

**Table 3.4. Outcomes for the goalie of
the penalty shootout game.**

Again, there is no minimax solution for a single game. The tree for the penalty-kicking game with only two options is the same as the choosing game above, but now it gets more interesting.

---

* Remember these models are simplified versions of reality. In a real penalty shoot-out, the penalty taker might also miss the goal, and the goalie could go the same way as the ball, but fumble the catch and let the ball through. The model can be made more sophisticated to deal with these options – this is left as an exercise for the reader.

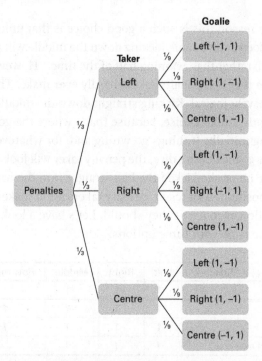

**Figure 3.3. Tree for penalties, with outcomes
for taker, goalie in brackets.**

If the taker and the goalie choose randomly between the
three options, the taker gets six out of nine goals because
there are two locations where the goalie isn't. Making a random selection results in a positive outcome for the penalty
taker (though he or she will be embarrassed one-third of
the time).

Of course, if the goalie always dives, the minimax strategy would not be best for the penalty taker, who should
instead always aim for the centre – assuming no misses, they
would get a goal every time. The best strategy in practice,
though, is likely to be dynamic. The penalty taker would

start by kicking for the centre all the time. Goalies would gain information on this strategy as the ball continues to slam into the net and would start to stay in the centre all the time. After some oscillation, the outcome would eventually settle down into the minimax solution of the penalty taker choosing randomly between the three options (provided both the penalty taker and the goalie are willing to take the abuse that will arise from staying put or kicking to the centre).

Mix master

In real-world penalty shootouts, logic rarely prevails. However, if we are in a circumstance where the players have the time and the mental capability to think through a strategy, then it is possible to have a process that will generate the correct balance for a mixed strategy. A starting point is to establish that there isn't a clear single strategy with a saddle point as was the case with the cake-cutting game. Assuming that there isn't, then it's possible to calculate what proportion of choices each player should select to produce a mixed-strategy minimax outcome.

Let's look at the colour-matching game again, but with different payouts.

Player 1 ↓ Player 2 →	Blue	Red	Row minima
Blue	–£2	£3	–£2
Red	£1	–£2	–£2
Column maxima	£1	£3	

Table 3.5. Outcomes for player 1 in the colour-matching game with different payouts.

First, as always, we check for a saddle point. There is no max-imin for player 1, so a mixed strategy is necessary. Imagine first both players went for a 50:50 mixed strategy – tossing a coin. The expected outcome over time for player 1 would be $(-2+3+1-2)/4 = 0$. Player 2, of course, has the same outcome, as this is a zero-sum game.

The mechanism for working out the best mix of available strategies is relatively simple, though it feels a bit odd. If you don't want to work through the arithmetic you can skip to the next section, 'Line of Defence', starting on page 58. (If you do decide to skip forward, the result of the calculation is that should both players play rationally, the best player 1 can do is to lose £0.125.)

Let's see the mechanism in action first, then get a feel for why it works. For each player, we subtract the second value from the first in the relevant row or column, then use the absolute value* of the result as the ratio for the *other* strategy. That sounds more complicated than it actually is: taking a specific example will make it clearer.

Player 1 subtracts the second value in the blue strategy row from the first, which is $-2 - 3 = -5$. They do the same for the red strategy row: $1 - -2 = 3$. The ratio of blue strategy to red strategy is the ratio of the red total to the blue total (or rather the absolute values of each, so 5 instead of -5). So, player 1 should play the red strategy five times for every three times he or she plays blue.

Player 2, meanwhile, subtracts the second value in the blue strategy column from the first: $-2 -1 = -3$, with an absolute value of 3 (taking the absolute value does away with

---

* The absolute value is the number without any sign – so the absolute value of both 3 and -3, say, is 3.

the problem that we are using the outcomes for player 1 in both cases). The red strategy column gives 3 – –2 = 5. So, player 2 should play the blue strategy five times for every three times they play the red strategy.

The minimax theorem says that if one player plays their best strategy, they will get at least the value that arises from the 'game value' which is the figure that comes up if both play their best strategy, though they could do even better if the other player does not play their best strategy. The reason this mechanism works is that it calculates the value where the player is indifferent between the two choices available, and so should be happy selecting probabilistically.

In this game, that game value, for player 1, is calculated using the formula:

$$(P_{1B} \times P_{2B} \times O_{BB}) + (P_{1R} \times P_{2B} \times O_{RB})$$
$$+ (P_{1B} \times P_{2R} \times O_{BR}) + (P_{1R} \times P_{2R} \times O_{RR})$$

where P_{XY} is the probability of Player X choosing option Y, and O_{XY} is the outcome (for player 1) of player 1 choosing X and player 2 choosing Y.

$$(\tfrac{3}{8} \times \tfrac{5}{8} \times -2) + (\tfrac{5}{8} \times \tfrac{5}{8} \times 1) + (\tfrac{3}{8} \times \tfrac{3}{8} \times 3)$$
$$+ (\tfrac{5}{8} \times \tfrac{3}{8} \times -2) = -0.125.$$

This means that game is biased against player 1. As long as player 2 plays their optimum strategy, player 1 will lose £0.125 each game, whatever their strategy. If, for example, player 1 used the 50:50 mixed strategy, but player 2 stuck to the minimax strategy, the outcome for player 1 would be:

$$(\tfrac{1}{2} \times \tfrac{5}{8} \times -2) + (\tfrac{1}{2} \times \tfrac{5}{8} \times 1) + (\tfrac{1}{2} \times \tfrac{3}{8} \times 3)$$
$$+ (\tfrac{1}{2} \times \tfrac{3}{8} \times -2) = -0.125.$$

If player 1 plays their optimum game, the most they will lose is £0.125. However, bear in mind that this is worse for player 1 than both players making random choices. If player 1 knew that player 2 was going to choose randomly, then player 1 would be best doing so as well. Similarly, if player 2 were silly enough to always choose the red strategy, all player 1 needs to do to clean up is to stick to blue.

Line of defence

The example above seems quite arbitrary, because it's hard to imagine a real-world situation given only a set of outcome values. But making the options closer to reality can make the power of game theory more obvious.

Let's imagine a coffee chain is interested in moving into a city where it hasn't previously operated. A local coffee company already has two outlets in the only two locations where coffee shops are worth operating. It's down to the local council to decide whether or not the current owner can keep operating in these locations or whether they can be forced to sell them to the chain.

The current owner can choose to defend either of his shops by offering a bribe to the district councillor – but they can only choose one shop to defend. Similarly, the chain can force the purchase of one shop, as long it is not defended. One shop is much bigger and worth three times as much as the other. Here's the game in numbers (see Table 3.5).

There is no saddle point, so the coffee shop owners need to adopt a mixed strategy. Working through the numbers we discover that the value of the game is 3.25, with the local defending his big shop three times out of four,

Local ↓ | Chain →	Big shop	Small shop	*Row minima*
Big shop	4	3	*3*
Small shop	1	4	*1*
Column maxima	4	4	

Table 3.5. Relative incomes for the local shop owner as a result of the coffee shop game.

while the chain should attack the small shop three times out of four.

It might seem counterintuitive for the chain to be most likely to aim to take over the small shop but, thinking through the strategies, the locals are more likely to defend their most precious asset, so to win something, the chain should mostly go for the small shop. They should, however, try for the big shop one in four times to avoid being predictable. As always, if one player always adopted the same strategy and this information become public, then opposing them would be easy.

Burgers or ice cream?

Another example where game theory can have a real application for a business is where conditions determine sales. Let's imagine a mobile catering firm sells both burgers and ice cream. Not surprisingly, on a cold day, burgers are more popular than ices, while on a hot day, ice cream is the big seller. Unfortunately, due to having a dodgy freezer, any food the caterer doesn't use on the day has to be thrown away. As our seller has been operating some years, he has a feel for the potential sales he can make. When it's hot, he sells around

200 ice creams and 50 burgers. When it's cold, he sells about 20 ice creams and 150 burgers. Burgers cost £2 and sell for £5; ice creams cost 50p and sell for £1.50.

With this information, he can produce a game plan based on profit (or loss) from his expected outlay and earnings. As can be seen from the table, the problems come when he buys stock for hot weather but tries to sell it in cold weather, or vice versa.

Weather: Expected ↓ | Actual →	Hot	Cold	Row minima
Hot	£350	–£30	–£30
Cold	–£120	£470	–£120
Column maxima	£350	£470	

Table 3.6. Profit/loss for the owner dependent on stock purchase and the weather.

There is no saddle point, so the caterer needs a mixed strategy. Crunching the numbers gives him a ratio of buying hot 59 times to 38 cold – roughly 3:2. There are two ways he could play this – either to go for the odds and select randomly between hot and cold such that three out of five times he picks hot, or to go for a compromise of buying ⅗ of what he would buy for a hot day and ⅖ of the cold day purchase. The result should be a steady profit of just under £166 (the value of the game).

In reality, of course, the chances are that our caterer would have access to a weather forecast, so could do considerably better. In this circumstance the game would be based on how often the forecast was right and how often it was sufficiently wrong to change purchasing habits – but the same broad principle could be applied.

Is this for real?

Although calculating the values for mixed strategies works mathematically, it has been suggested that this demonstrates the distance between reality and game theory, as people making normal decisions are not going to base their choice on this kind of complex calculation. This seems to imply that game theory does not provide an effective match to reality.

There are two reasons why this argument is faulty. One is that game theory does not claim to model human behaviour, but rather it models the decision process, giving us a better understanding of the choices available in the decision. This enables the decision-makers to have a better chance of making a good decision.

The second reason is to recall the nature of models. Science uses them all the time because they give useful insights, but they often involve extreme simplification. This is illustrated well by an old science joke. A dietician, a geneticist and a physicist are each set the task of deciding how to win a horse race. The dietician describes the particular diet the horse should be fed for the weeks leading up to the race. The geneticist tells the stable how to breed selectively for certain capabilities – and perhaps even how to use CRISPR, a tool to edit genes, to breed a winner. Meanwhile, the physicist scratches her head. 'Let's assume the horse is a sphere...' she says.

The joke is meant to illustrate how much physics tends to simplify reality to apply its mathematical models; but, in reality, all three are simplifying things, in part because they are making use of the 'law of the instrument'. This is a tendency to produce a solution that depends on the tool available. The name of the law comes from American philosopher Abraham Kaplan who first referred to it in 1962, and is often linked to a quote from American psychologist Abraham Maslow, who wrote in 1966: 'I suppose it is tempting, if the only tool you have is a hammer, to treat everything as if it were a nail.' Of course, the idea is much older – in the UK, for example, a hammer has been jokingly referred to as a Birmingham screwdriver for well over a century.*

When the dietician, the geneticist and the physicist come up with the strategies to develop a winning racehorse, each uses the simplified model developed by their own discipline: in reality the quality of the racehorse is dependent on all three inputs and many more as well. While game theory's solutions are often as limited as these scientific models, this does not mean that they are without value.

Although there are examples where the values that go into the game table are known, often they will not be certain. In the caterer example above, while he knows the cost of purchases, he does not know what his actual sales will be. The numbers are an estimate. But this is true every time someone makes a budget, or any other forecast of the future. It will almost certainly be wrong – but we have to make a decision, and using a best estimate is preferable to not trying at all.

What are the chances?

We've already seen in the matching game an example of a game where a mixed strategy applies because there isn't

---

* There are local variants of the 'Birmingham screwdriver' too. In the Lancashire town of Rochdale where I was born, for example, a hammer used to be known as a 'Heywood screwdriver', to be derogatory about a nearby town.

an appropriate minimax strategy from a single play. In that game, the outcomes are the same each time the game is played – but real life is rarely so consistent. And this can result in apparently baffling game tables like this one:

Player 1 ↓ Player 2 →	Blue	Red
Blue	£0	£0
Red	£0	£0

Table 3.7. Expected outcomes in a game with varied probabilities.

At first sight this is a totally pointless game where there is no point in playing, because whatever you do, no one wins anything. You might as well go and watch the TV instead. But hidden beneath those zeros is a far more interesting probabilistic outcome.

Imagine, for example, that there is a game where each player throws a die in turn. The outcomes from that throw are as follows:

Throw	Outcome
1	£5
2	–£3
3	£10
4	–£8
5	£2
6	–£6

Table 3.8. Player 1's outcomes for each possible throw of a die.

Here a positive figure means that the thrower wins that off the other player, while a negative number means that the thrower has to pay that amount to the other player. There is a one in six chance of each of the possible outcomes, so the expected win from repeated games is the sum of all the outcomes, divided by six. In this case, the outcomes have been devised so that their sum equals zero. So, in the long run, the expected outcome is that neither player comes out on top.

Bringing probability into the game has made it more interesting than the averaged outcome of winning and losing implies. A player could win as much as £10 or lose up to £8 on a throw (and the reverse on a throw by the other player). However, in game theory terms this is still not an interesting game because there is only one approach available – throw the die and win or lose accordingly. There is no decision to make, so there can be no real strategy.

It doesn't take much effort, though, to take the mechanism and turn it into a game where there is more than one option (here 'red' and 'blue'), so we can get to the outcome table above. Note that there is no need for row minima and column maxima here, as it is clear that there is no saddle point.

Red strategy		Blue strategy	
Throw	Outcome	Throw	Outcome
1	£5	1	–£10
2	–£3	2	£8
3	£10	3	–£5
4	–£8	4	£12
5	£2	5	–£7
6	–£6	6	£2

Table 3.9. Player 1's outcomes on each possible throw of a die with differentiated strategies.

The outcome table still provides equal outcomes of £0 whichever strategy is chosen, but now utility comes into play. A player that is less sensitive to losses might opt for the blue strategy, while a more conservative player would prefer the red.

Haggling horrors

Probabilistic games are often a lot closer to many of our real-life experiences, because it can be difficult to apply a definitive numerical outcome to an interaction ahead of time. One clear example where it feels like many of us would benefit from applying game theory to a decision is in purchasing something. We tend to be fixated on a list price, but all prices are negotiable. Goods and services are only worth what someone will pay for them. Culturally, though, many from a Western background have ceased to understand the importance of haggling.

Although haggling is possible in all transactions, three circumstances where it is still relatively common are in private sales; when a time-limited contract runs out; and when negotiating a freelance contract. There is reasonable evidence that even, for example, supermarkets will accept haggling, but it is significantly harder to do so in such circumstances as it is usually necessary to bypass shopfloor staff and reach management, and because of the need to circumvent normal sales procedures, meaning that the supermarket is only likely to be interested in making the effort for high-value sales.

The problem with taking a numerical game theory approach to these transactions is that initially we have no information on the other player's strategy. So, for example, if you are selling something or making a pitch for a freelance contract, it is unlikely that you will know how a potential buyer will react to a price. It's useful to survey any existing prices, which, thanks to the internet, is certainly relatively easy to do for a private sale, such as a house sale, where websites provide sale prices on similar properties. Or online marketplaces can give a feel for the prices at which others are offering similar products. Freelancers are more in the dark – it is here that associations can be helpful, either in providing a mechanism to contact peers, or providing a guide. So, for example, the Society of Authors gives a suggested minimum for a speaking fee.

As we will see (page 133), the favoured game theoretic tool for revealing price information is often an auction. Sites such as eBay make it possible to sell via auction and discover the values that buyers place on products. Arguably the best approach here, if multiple products are available, is to use one or more auctions to set a band for the pricing, then try direct sales at, say, 30–50 per cent above the price reached at auction, bringing the value down over time and being prepared to negotiate. Again, sites such as eBay now provide quite sophisticated tools for doing this.

As an example, I recently sold two copies of a rare book, known to be in high demand on eBay. The first, sold at auction, raised £100. This gave a broad feel for the range of interest (and provided information on the size of the potential pool of buyers, judged from the number who followed the auction). The second book was initially listed at £140, then brought down to £130. A buyer offered £120. I counter-offered £125, which was accepted.

A similar approach can almost always be employed with contract renewal and large-ticket purchases. When, say, buying a house or a car, or renewing an insurance or broadband contract, all too often purchasers go with the stated price, but it is always possible to negotiate a better one. There is a (complicated) mathematical approach to such bargaining games (see page 84), but it is not a practical tool because of a lack of information.

Even so, the principles of game theory, specifically having a conscious strategy and being as aware of the other player's strategy as possible, still hold. So, for example, the strategy of a broadband provider at contract renewal is to increase the rate payable over the previous contract period, but to be prepared to reduce that rate on threat of switching to an alternative, as it is harder to gain a new customer than to retain an existing one. Being aware of this, the customer should always threaten to leave, rather than simply accept the new contract price.

Casting lots

An unusual one-sided game theory decision that millions make each week involves the concept that gave birth to the mathematical theory of probability in the first place: gambling. Lotteries provide an excellent example as they are far more widely played than traditional gambling games. It may seem that, just as in Table 3.7 above, the fixed set of payouts from a lottery mean that there can be no strategy involved. The player might as well just choose numbers at random. However, in practice, the lottery player usually has several strategic choices to make given a certain amount of stake money, which can benefit from the application of game theory. Let's look at two such choices: which numbers to pick and which of two lottery games to place a stake on.

One of probability's more confusing aspects can result in the application of a bad strategy on picking numbers. Over time, because each number has the same chance of being chosen, we expect each possible number to be drawn roughly the same number of times. With the number of draws taking place, however, it would be surprising if all the numbers came up *exactly* the same number of times – and this is reflected in the statistics.

At the time of writing, the UK's national lottery Lotto game uses 59 balls to generate six main numbers. If we look back over the last year, the most common balls to be drawn were 52, 49 and 28, each drawn seventeen times. We would expect over time that each ball would come up in 1.69 per cent of draws, whereas each of these balls has come up in 2.7 per cent of cases.

At the bottom end of the ranking, we have three numbers (30, 43 and 58) that have only been drawn six times, with a single outlier, 41, only picked four times. These are 1 per cent and 0.6 per cent of the draws respectively. There are two ways to be guided by these numbers that are based on fallacy, and one where there is a possible contribution to a useful strategy.

The most extreme misguided strategy is to think that some balls are 'lucky' and so more likely to turn up than others. Here, the strategy would be to choose the most frequently appearing balls: 52, 49, 28, followed by the next most frequent, 1, 33, and 50. This strategy appeals to nothing more than magic. An apparently more logical approach is to choose the balls that haven't appeared very often, because it seems reasonable that over time the numbers picked will even out. This approach would see us choosing the numbers, say, of 41, 58, 43, 30, 25 and 40. (The last two of these choices could be any pair from 25, 40, 14 and 12, each of which has turned up seven times.)

Although this second strategy feels more reasonable, it is also based on a fallacy. To see why, it's easier to temporarily simplify the probabilities involved to a fair coin toss where there is a 50:50 chance of heads or tails. Say we've thrown four heads in a row. Should you choose heads or tails as a strategy for predicting the outcome of a fifth throw? The correct answer is that each has the same likelihood of being thrown. The coin has no memory. The fact that there have been four heads in a row beforehand has no influence on what is thrown next. Unless that string of heads (which will occur on average one in sixteen times) was due to having a biased coin, it says nothing about what will be thrown next.

Even though we understand that the coin has no memory, this feels counterintuitive. We might believe that after thousands of throws we would expect the number of heads and tails to be pretty much the same. That initial bias can only be 'fixed' by there being more tails later on. But that is not what happens. It is entirely possible after thousands of throws for there to be a big difference between the number of heads and the number of tails. The *average* proportion of heads and tails will tend towards 0.5, but the difference between the number of heads and the number of tails could continue to grow. There is no force of nature trying to fix a run of the same throw. (Also, bearing in mind there is a one-in-sixteen chance that four throws in a row will all be the same, either four heads or four tails, if you are tossing a coin thousands of times there will be plenty of examples of such runs.)

Going back to the lottery, exactly the same applies. The lottery draw machine does not have a memory of the balls that have previously been drawn. Each draw starts from scratch and cannot be influenced by what has happened previously. These statistics are meaningless as a predictor of what will happen next.

What, then, is the possibly useful strategy arising from this information? Many lotteries divide the main jackpot among winning tickets. Although the lower-value prizes are usually fixed, this means that if, say, two people win the jackpot, each only wins half as much as would a single winner. As at least some players *do* follow the frequency with which balls are picked, the useful strategy emerging from this rule is to avoid numbers that others are more likely to pick. By trying to make use of numbers that have either frequently or infrequently come up in the draw, players will select balls that are more likely to be drawn by others than if they selected at random. The strategy, then, would be to avoid numbers that have either been frequently or infrequently drawn.

In practice, though, the players who ignore draw frequency don't all pick numbers at random, which adds a second opportunity to hone a strategy that will improve your chances of not sharing the jackpot should you win it. (It should be stressed that winning the big prize is still very unlikely. The chances of winning the UK Lotto game jackpot are around one in 45 million.) Perhaps the most common ways used to choose a number are to use a house number or a significant date.

If a road has fewer numbers in it than balls in the lottery, then there will be some bias towards the lower numbers. This suggests there is no harm going for the larger numbers, all other things being equal. If a road has more numbers than there are balls it's less clear what will happen if someone, say, lives in house number 101 or 199.

A stronger influence is significant dates. There is something many find attractive in using, for example, their children's birthdays. Picking a day of the month limits the number to under 32, while using a month keeps it under thirteen. This suggests numbers from 32 up are less likely to result in a shared jackpot. The year is more of a fluid choice. Most people will be in the 0–80 age range. At the time of writing in 2021, this means we will see most birth years being between 1941 and 2021, so a relatively low number of players will choose numbers between 22 and 40 based on year – however, lotteries tend not to go above 60, so years will be used less frequently than day and month numbers.

A final consideration is sequential numbers. Although some like the attractive pattern of a choice like 1, 2, 3, 4, 5, 6 – and such runs are chosen more often than you might expect – many will feel this is less likely to occur than, say, 5, 17, 23, 25, 39, 52. In fact, both have exactly the same chance of being picked, so having at least some sequential numbers is probably also beneficial to avoid a shared jackpot.

It ought to be stressed that this guidance is useless in lottery games where a fixed sum is paid to each jackpot winner. In such circumstances, the set of numbers chosen is unimportant, but the game chosen can give the player a choice of strategies. In the UK at the moment, for example, two games with similar stakes and a fixed jackpot (apart from under unlikely circumstances) are Thunderball and Set for Life. Each has a different set of possible payouts, which makes it difficult to make an informed choice – but by comparing the outcome of the different strategies it is possible to make the best decision.

Below are the payout tables for the two games, which make use of the concept of expected value. As we have seen, this is found by multiplying the prize by the chance of getting the prize. The expected value tells us how to compare the games long term. However, it's also worth remembering that we also need to consider how important to us is the amount we could win and lose: the utility. For example, a game where you have a one-in-ten chance of winning £1 million has an expected value of £100,000 – so if this were small change to you, you should play the game, even if it cost, say, £99,000 to play. On average, you would benefit by £1,000 times the number of games played. But for most of us, £99,000 would not be an amount we could afford to lose even once.

So, those tables. First, Thunderball:

Odds	Prize	Expected value
1 in 8,060,598	£500,000.00	£0.062030137
1 in 620,046	£5,000.00	£0.008063918
1 in 47,416	£250.00	£0.005272482
1 in 3,648	£100.00	£0.027412281
1 in 1,437	£20.00	£0.013917884
1 in 111	£10.00	£0.09009009
1 in 135	£10.00	£0.074074074
1 in 35	£5.00	£0.142857143
1 in 29	£3.00	£0.103448276
TOTAL VALUE		**£0.527166285**

Table 3.10. Payouts and odds in Thunderball game.

And then Set for Life:

Odds	Prize	Expected value
1 in 15,339,390	£3,600,000	£0.23468991
1 in 1,704,377	£120,000	£0.07040696
1 in 73,045	£250	£0.00342255
1 in 8,116	£50	£0.00616067
1 in 1,782	£30	£0.01683502
1 in 198	£20	£0.1010101
1 in 134	£10	£0.07462687
1 in 15	£5	£0.33333333
TOTAL VALUE		**£0.8404854**

Table 3.11. Payouts and odds in Set for Life game.

Here the 'odds' refer to the chance of winning that particular prize, so odds of one in 35 mean you have that chance (which corresponds to 0.029 or 2.9 per cent) of winning £5 in Thunderball. Although the total value of Set for Life is higher, Thunderball costs £1 to enter while Set for Life costs £1.50, so the best comparison would be the outcome of entering Thunderball three times compared with entering Set for Life twice.

Going on averages alone, we would expect a return of around £1.58 from our three entries to Thunderball and £1.68 from the two entries to Set for Life – very similar outcomes, though Set for Life has a small edge. But going into the detail of the expected values provides some interesting extra detail to contribute to the choice of strategy.

If we exclude the expected return from less likely outcomes – say, beyond one in 100,000 – we can see which strategy is best depending on our attitude to risk. In the table below, we look at expected returns on a £3 investment ignoring the low-probability paybacks.

	Thunderball	Set for Life
Better than 1 in 100,000 chance	£1.37	£1.07
Better than 1 in 10,000 chance	£1.36	£1.06
Better than 1 in 1,000 chance	£1.23	£1.01
Better than 1 in 100 chance	£0.74	£0.67
Better than 1 in 30 chance	£0.31	£0.67

Table 3.12. Expected returns within certain risk levels from the two games.

We have seen that if you are happy to take large risks and look at the total expected value, playing Set for Life is a slightly better strategy. But if you're a medium risk taker, limiting your risk to the values set out in the table above, Thunderball represents a better strategy. Finally, if you like to minimise risk, because Set for Life has a significantly better chance of winning the lowest amount, it suddenly dominates the strategies. For many players who are looking for a mix of risk outcomes, the best option would be to play a mix of the two games.

This is a good example of the way that game theory is rarely going to tell you exactly *what* to do, but can be helpful as a way of examining your different strategies and their impacts.

That game

Arguably, the most famous example of a 'game' specific to game theory is the prisoner's dilemma, not only because it stretches our moral considerations of choice but because it was used by some (including the hawkish John von Neumann) to argue for the necessity of making a pre-emptive nuclear strike during the Cold War. This game was devised in 1950 by Merrill Flood and Melvin Dresher, who were working at RAND Corporation.

RAND would play a big part in the advancement in America of the kind of strategic thinking encompassed by game theory. The corporation emerged from a 1945 project by the Douglas Aircraft Company to provide research and development (R and D – hence RAND) for the US Armed Forces. As well as smaller-scale decisions, the organisation developed strategy analyses on the potential for and outcomes of intercontinental nuclear war. RAND was spun off as a non-profit organisation in 1948, and has gone on to provide a wide range of strategic studies. The institute used von Neumann as a consultant for a number of years.

In the original version of the dilemma, small sums of money were involved, but a variant of it in which two prisoners are facing a period in jail gave it the familiar name and a particularly visceral impact in its apparent implications for ethics. In this game, each prisoner is offered the opportunity to give evidence against the other. If both give evidence, both go down for a long stretch. If one gives evidence but the other won't, the one giving evidence goes free and the other has an even longer sentence. And if neither gives evidence, they both receive a short sentence in jail.

If, for example, the sentence when both give evidence is seven years, the sentence when neither gives evidence is one year, and the sentence when only one gives evidence is ten years, then the outcome table will look as shown in Table 3.13.

Note that von Neumann's minimax theorem doesn't apply here, as this game isn't zero sum. The amount of time one person spends in jail is not numerically opposite the time

Prisoner 1 ↓ | Prisoner 2 →	Give evidence	Withhold evidence
Give evidence	7 7	10 0
Withhold evidence	0 10	1 1

Table 3.13. Prisoner's dilemma outcomes (prisoner 1 bottom left values, prisoner 2 top right values).

the other is incarcerated. To show the unbalanced outcomes for the two players, prisoner 1's outcomes are in the bottom left of each cell above and prisoner 2's in the top right.

What isn't immediately clear here is what is the best outcome. If we look at the common good – minimising their combined sentence – it is clear that both should refuse to give evidence, resulting in their only getting two years in prison between them. However, if one prisoner knew that the other was definitely not going to give evidence, they could get their best personal outcome by giving evidence – getting off without any time in jail (but with the other person's long sentence on their conscience). If both give evidence, they get what is collectively the worst possible outcome of a total of fourteen years in jail between them.

What is confusing about the dilemma is that if we take either player, there appears to be only one rational choice for them to make. Say we look at prisoner 1. In the case that the other player gives evidence, prisoner 1 would be better off giving evidence too (getting a seven-year stretch, rather than ten). And if prisoner 2 withholds evidence, prisoner 1 would still be better off giving evidence, as prisoner 1 would then be released immediately rather than spending a year in jail. It doesn't matter what prisoner 2 does, prisoner 1 is better off giving evidence.

The game is symmetrical, so exactly the same logic applies when considering prisoner 2's choice. Whatever prisoner 1 decides to do, prisoner 2 is better off giving evidence. Yet despite this clear, rational outcome that both individuals should give evidence, if the two prisoners can cooperate and both withhold evidence, that will achieve the best outcome overall.

It's worth emphasising this point, as it's what makes the prisoner's dilemma so mind-boggling. Whatever the other prisoner does, it is better for a prisoner to give evidence. Yet if they could cooperate and both withhold evidence, they would do far better than if they take that apparently inevitable rational choice of both giving evidence.

The key phrase there is 'yet if they could cooperate'. Let's turn the game round to a positive rather than a negative outcome, which can make the benefit of cooperation clearer. Imagine a game with up to £10 available to be given away. Each player can either give the other player £5 from the pot, or take £1 for themselves.

Player 1 ↓ | Player 2 →	Give		Take	
		5		6
Give	5		0	
		0		1
Take	6		1	

Table 3.14. Financial outcomes for the players of a positive-outcome prisoner's dilemma.

Again, in terms of personal reward you are always better off by taking. If the other person takes, you get £1 rather than 0. If the other person gives, you get £6 rather than £5. However, it is absolutely clear that taken as a whole, the give/give outcome of getting £5 each is far better than the take/take outcome of getting £1 each. Note that the cooperative outcome can be made the rational choice for the players with a slightly different version of the payout. Here, rather than receiving individual winnings, the pair split the total winnings between them. If that were the case, both giving nets them £5 each, one giving and one taking nets £3 each and both taking nets £1 each, making cooperation the clear choice of strategy.

Going MAD

Although the prisoner's dilemma was not produced with real-world applications in mind, it was devised at RAND, where a major focus was the development of strategies for the use of nuclear weapons. It's not surprising, then, that the dilemma was seen as a potential model for two stages of nuclear armament strategies.

The first consideration was to support the concept of nuclear deterrence. Leaving aside the difficulties of acquiring them, countries have the choice of either having or not having nuclear weapons. If we imagine, say America and the USSR (as the superpowers were back then) facing off over the development of the hydrogen bomb, the ideal collaborative outcome would be that nobody builds the bomb and everyone is safe from this most extreme source of nuclear devastation.

However, from the viewpoint of either player, whatever the other does, it is preferable to build the bomb. If they both end up building it, it means they have a deterrent; if only one side builds it, they have the upper hand. As a result, both sides inevitably build the bomb, heading for the same equilibrium choice. This is the doctrine that became known as 'mutually assured destruction' or MAD.

The second step takes things further. Exactly the same argument can be used for undertaking a pre-emptive nuclear strike on the opposing player. If they both strike, then at least the destruction is punished. If only one side strikes, that side has rid itself of its enemy. Thankfully, this game was not taken to this rational conclusion. The cooperative choice wins, not because of its rationality in pure game theoretic terms, but because of its recognition of a combined benefit to humanity, just as the choice to cooperate in the basic prisoner's dilemma game results in a clear benefit for the two prisoners over both giving evidence.

One oddity here is that seeing nuclear stand-off as a prisoner's dilemma means that it is not being treated as a zero-sum game. Yet the more hawkish among those involved on both sides may have been likely to consider the situation as such. If nuclear warfare were a zero-sum game, if, say the USSR was destroyed, from the US viewpoint, this would be pure benefit. One of the reasons that the prisoner's dilemma is so interesting is that it can accommodate a less polarised view and still come to an apparently rational assumption that lacks humanity.

To get a better understanding of the prisoner's dilemma, we need to introduce the second key player in the development of game theory, John Nash.

REACHING EQUILIBRIUM 4

Meet John Nash

John Nash has become a familiar name thanks to the Hollywood movie based on his life, *A Beautiful Mind*. Nash was born in Bluefield, West Virginia in 1928. His father, John Forbes Nash Snr, was an electrical engineer who spent most of his life working for the Appalachian Power Company, while Nash's mother, Margaret Virginia Martin, was a teacher until she married, at which point (as was common in the 1920s) she was no longer allowed to continue in her career.

Nash was an introverted child who generally preferred to play alone indoors. Probably because of a lack of social skills, he was initially labelled a low achiever at school (including in mathematics). As had been the case for Albert Einstein – much later a colleague at the Institute for Advanced Study – Nash seemed happier learning from books than in school. Unlike Einstein, though, he also had a passion for experimentation, in both electrics and chemistry.

It soon became clear that Nash's apparent failings in

mathematics were not so much his as those of his teachers. The school curriculum followed set approaches to solving mathematical problems, considering any other approach to be incorrect. Nash found that he could cut through the often laborious taught mechanisms, devising his own, more elegant solutions, something the teachers weren't equipped to deal with.

We don't need Nash's level of mathematical expertise to see how this can be the case. There was a social media flurry in 2015 when a child in America lost marks in a test. The students were asked to work out 5×3 using repeated addition. The child in question used $5 + 5 + 5$ rather than $3 + 3 + 3 + 3 + 3$ and was told that he was wrong. According to the teacher, this was because 5×3 means 'five lots of three'.

However, there is a problem with this logic – it entirely depends on the teacher's somewhat childish English syntax. This is because 5×3 could also be said to represent 'five multiplied by three'. This means $5 + 5 + 5$ because multiplication involves making use of the 5 three times. In the end, multiplication is 'commutative', meaning that $5 \times 3 = 3 \times 5$. Either approach is entirely correct. On a more dramatic scale, Nash would approach a mathematics problem in a totally different way to that envisaged by his teachers, causing them to mark him down incorrectly.

As the Second World War came to an end, Nash travelled to Pittsburgh, Pennsylvania to attend the Carnegie Institute of Technology. Under pressure from his father, he applied for chemical engineering, but soon spent his time solely on mathematics. It was here, with teachers who really understood the subject, that Nash's original thinking and mathematical ability started to shine through. Academically he began to excel, though remaining socially inept: he struggled to get on with his peers. From Pittsburgh, he moved on to Princeton, which had an unusually strong mathematics department, in part due to strong connections with Europe.

Nash flourished in the results-oriented atmosphere that had been engendered at Princeton, intellectually aggressive and taking no prisoners with his competitive attitude. He attended few lectures; nor, by this stage, did he read much – a lot of his time was simply spent thinking about and working on mathematical problems. In this academic hothouse, he strayed from his social isolation, always interested in discussing problems, provided he thought that the conversation was with someone of worthwhile talent. Otherwise, he could be instantly dismissive.

Board games were popular in the Princeton maths department, and Nash was an enthusiastic player, notably of Kriegspiel, a sort of hybrid of chess and Battleships, where a player can only see his or her own pieces. He also devised a game of his own, known at Princeton simply as 'Nash' or 'John'. A variant of it was independently invented a few years before in Holland, and it would be commercialised as Hex. The game took place on a diamond-shaped board with hexagonal 'squares' (the Princeton board was devised by a fellow student), which players occupied with either black or white stones, like Go stones.

The winner of John was the first to join opposite edges of the board (each had an allocated pair of edges) with an unbroken line of stones. In devising the game, Nash had come up with a two-player, zero-sum game which, like chess, had 'perfect information'. Having perfect information means that there is nothing hidden from any player. In chess, for example, you know how the board was first set up and what every move has been since. In most card games, by comparison, you do not have perfect information as you do not know what is in the other players' hands. In Nash's game, in principle, with a perfect strategy the first player to play could always win – though Nash made it clear that he didn't know what that strategy was. Simpler than chess, the game still had too many possible moves to have an explicitly known complete strategy.

It was John von Neumann who got Nash interested in the concept of game theory. At first sight, Nash was not an ideal match for the theory. He regarded much applied mathematics as worthless – and the mathematics of games is indeed largely trivial from a mathematician's viewpoint. But Nash was swept up in von Neumann's charismatic wake. Here was a man who was an academic but who was also deeply engaged in the worlds of business and politics, and whose influence was immense. Von Neumann had a glamour that was rarely associated with mathematicians.

Along with many of his peers, Nash immersed himself in von Neumann and Morgenstern's heavyweight book *Theory of Games and Economic Behavior* (see page 123). This is strong on two-player, zero-sum games – games of pure conflict – but these would prove to be relatively rare as models of human interaction. Von Neumann was less successful when dealing with non-zero-sum games. Nash saw an opening and picked up on what can be seen of one of the fundamental games of the economic sphere – bargaining. This formed the basis of Nash's first academic paper.

Nash's approach is rather different to the games we have seen so far, combining as it does the alternatives that those involved in the bargain have and the potential benefits they would gain from making an agreement, making use of the relative utility of the components of the deal to the players.

He then used a graphical approach to identify the combined maximum utility for the players (we will encounter a variant of this method on page 100).

After passing his exams, Nash had to put together a thesis topic. He suggested an idea to von Neumann, but von Neumann regarded it as insignificant. However, Nash pressed on, effectively discarding von Neumann as a stepping stone to his future. Nash's idea was a way to go beyond minimax to provide a solution that would apply both to non-zero-sum games and those with more than two players. The concept, which we will explore in the next section, would become known as the Nash equilibrium and would form Nash's greatest contribution to game theory. At the time he was 21.

Nash made considerable advances in other areas of mathematics, particularly aspects of geometry and differential equations. For a considerable period of time, however, mental health issues wrecked his life. By the time he was 30, he was a professor at the Massachusetts Institute of Technology. At a time when he should have been at the peak of his influence, he began to worry his colleagues. On one occasion, for example, he told other lecturers in all seriousness that an article in the *New York Times* included an encrypted message from another galaxy.

For the next 30 or so years Nash suffered profound bouts of paranoid schizophrenia, was regularly hospitalised and was unable to make contributions to the work he loved. Some younger academics assumed he was dead, so entirely had he disappeared from the mathematical scene. It was only at the start of the 1990s, when Nash was in his 60s, that he (unusually for sufferers of this condition) made a significant spontaneous recovery, in time to receive the Nobel Prize in Economic Sciences in 1994. He lived on until 2015, dying at the age of 86.

Nash equilibria

The most significant contribution Nash would make to game theory was to describe an outcome known as a Nash equilibrium. Imagine, for example, a game where each player can choose red or blue. Player 1 chooses red and player 2 chooses blue. This solution is a Nash equilibrium if things would not be better for player 1 by choosing blue, and things would not be better for player 2 by choosing red. In some games, such as the prisoner's dilemma, there is a single Nash equilibrium (in that case, reneging on each other), but there can be more than one. As this demonstrates, unlike von Neumann's minimax strategy, which only applies to zero-sum games, Nash equilibria are available for applications with more complex outcomes.

For example, consider a game where there is £5 available in a pot. Two players are given the choice of asking for £2 or £3. If they choose the same amount they get nothing, but if they choose different amounts they get what they ask for. Here, both outcomes where they win something is a Nash equilibrium, because in either case, each player is making the best decision given the choice of the other player.

Player 1 ↓ | Player 2 →	£2	£3
£2	0 0	3 2
£3	2 3	0 0

Table 4.1. Outcomes from the cash-choosing game – bold values represent Nash equilibria.

So, for example, if player 1 chooses £2, then £3 is the best strategy for player 2, while if player 1 chooses £3, then £2 is the best strategy for player 2. Because the same outcomes result from the best strategies for player 1, either of the highlighted outcomes above represents a Nash equilibrium.

This type of game used to be referred to as a 'battle of the sexes' game, where instead of money, the pay-offs represented activities a couple could do.

Player 1 ↓ Player 2 →		Movie		Meal	
Movie	3	2	0		0
Meal	0		0	2	3

Table 4.2. Outcomes from the battle of the sexes game – Nash equilibria in bold.

Here the aim is for the players to match, rather than choose something different. One of the two partners would prefer to see a film, while the other likes a meal better. Once again there are two Nash equilibria, in this case when the partners agree on an activity – but each partner prefers a different equilibrium. If the partners can discuss their strategies in advance, they can ensure that they achieve equilibrium. Because this is likely to be an ongoing situation, in practice they will tend to use a mixed strategy – perhaps taking turns to do the activity the other prefers.

However, the game gets trickier if it is being played as a one-off and if, for whatever reason, after the pair have decided to meet, the phone network goes down before they can agree on an activity. Admittedly this is an extremely artificial example, but in this case, the players have to rely on their guess of what the other would do – not always to their own advantage. We will return to this game a little later.

A game of chicken

A subtly different variant of this game has been a Hollywood teen movie favourite from *Rebel Without a Cause* to *Grease* (even though it rarely seems to have taken place in real life).

Player 1 ↓ Player 2 →	Swerve	Straight	*Row minima*
Swerve	5 5	8 **2**	*2*
Straight	2 **8**	0 0	*0*
Column maxima	**8**	**2**	

Table 4.3. Outcomes for chicken game – values are arbitrary, based on a combination of saving face and saving lives.

This is the game to which mathematician and philosopher Bertrand Russell, who used it as an alternative game model for nuclear deterrent, gave the evocative name of 'chicken'. It is usually portrayed as two cars driving towards each other down the middle of the road. If both carry straight on, they crash. If one swerves, they both survive, but the driver who did not give way gets the kudos. If both players swerve, they both survive and neither gets an advantage.

Like the previous game, there are two Nash equilibria for the combinations where one swerves and one carries straight on – so knowing the Nash equilibria gives no guide to a rational choice. Although this isn't a zero-sum game, it does have a minimax saddle point.* A player's maximum

---

* In the table above we see player 1's values. For player 2 we would need row maxima and column minima, which again will provide a saddle point at two.

minimum is two – the result of swerving (the minimum from going straight is zero). So the rational choice is for both players to swerve.

Russell's use of this game as a mathematical metaphor for nuclear stand-off came to life in October 1962 during the Cuban Missile Crisis, when the USSR began constructing nuclear missile sites on the Caribbean island of Cuba, located just over 100 miles off the US coast. This followed America's failed attempt to oust Cuba's communist leader Fidel Castro in the 1961 Bay of Pigs invasion. The crisis became an all-too-real game of chicken between the superpowers, with America's President Kennedy threatening a nuclear attack if the missiles were not removed. The fact that both superpowers survived emphasises that swerving was involved in the outcome. It's arguable that both sides swerved, though it seems likely that the US position was closer to a straight-line strategy.

It's also worth pointing out how the values assigned to outcomes have the potential to be misleading. The numbers we've arbitrarily assigned here, two and zero, don't seem very different. But in many modern cultures, the tendency would be to assign a much higher value to losing face and living than to keeping face and dying. This has not always been the case, and arguably is still not applicable in some cultures where death is considered preferable to a loss of honour.

This is important, as players might be tempted to go for a mixed strategy, swerving some of the time and carrying straight on at others. The higher the cost of a collision, the less frequently going straight on would come up in a mixed strategy. With a high weight on surviving over losing, the situation where both carry straight on and collide would be very unlikely to occur.

The bandwidth dilemma

Thankfully, very few of us have to make decisions involving deployment of nuclear weapons (or have been in a game of chicken) but we do experience the kind of game typified by the prisoner's dilemma where the 'rational' outcome of the Nash equilibrium involves both players suffering unnecessarily. What is missing in purely looking for a Nash equilibrium is whether or not the outcome is 'Pareto efficient'. A Pareto efficient solution is one where a player can't make him- or herself better off without making another player worse off. In the prisoner's dilemma, switching from the Nash equilibrium of both players reneging to both players cooperating makes both players better off. No one is worse off – it's Pareto efficient to cooperate.

A prisoner's dilemma interaction can take place where two individuals are sharing resources and either could dominate, but if both try to do so, the outcome is that both lose out. Perhaps the most widely studied modern example is wireless networks, where transmitters close to each other can interfere if they are too powerful, which reduces the available bandwidth for both.

If each transmitter reduces power, then both benefit from expanded bandwidth, but if only one reduces power, it loses out, as the other transmitter drowns it out. Left to their own devices, and given the power to act rationally, individual transmitters would opt for the high-power route, producing the Pareto-inefficient Nash equilibrium. But if the transmitters can interact, they can agree to lower power for mutual benefit. Unlike the basic prisoner's dilemma, this situation can be continuously monitored, so it is in the participants' interest not to renege on the agreement.

Understanding how the prisoner's dilemma works and its game theoretic implications can be a useful negotiating tool to emphasise the benefits of cooperation. It is easy when first learning of the prisoner's dilemma to think that it demonstrates the limitation of game theory, because the Nash equilibrium is not a desirable outcome – but this misses the point. It's easy to say 'What if everyone did that?', but this assumes that by suggesting the Nash equilibrium is the rational decision, game theory recommends doing this in all circumstances. The reality, that in some games the Nash equilibrium is not a desirable outcome, is even clearer in the wireless network problem's older cousin, the tragedy of the commons.

This is a situation where there is a limited shared resource, which would be sufficient for everyone if they all took a fair share. The 'commons' here historically referred to shared grazing land for animals. No one owned this land and individuals could choose how much of the resource they took. On an individual basis, they would do better if they took a little more than everyone else did. A few people doing this can be sustained – but once the majority start to do so, the whole system collapses. This, writ large is the problem of the Earth's natural resources. A few rich nations, representing a small percentage of the Earth's population, can take far more than their share of the available resources without it being a problem. But if the rest of the world starts to catch up – which is only fair – then there will be shortages.

From the game theoretic viewpoint, the rational thing is to take more than your fair share, even though this drives towards an equilibrium where everyone loses out. But bear in mind 'rational' here does not mean *sensible* from the viewpoint of the commons as a whole, only from the viewpoint of you, your family, your country at the immediate point of making the decision. As humans, we are able to look beyond that immediate short-term outcome and reason more widely, considering the benefits to society as a whole.

Forced cooperation

Sometimes where there is a real prisoner's dilemma, the players are forced into beneficial cooperation even when, left to their own devices, they might end up with an unsatisfactory Nash equilibrium. Such a situation happened to the benefit of major US tobacco companies in the 1970s. At the time, cigarette manufacturers spent a lot on advertising. Estimated annual profits (in millions of dollars, with inflation applied to 2021) for two of the majors dependent on advertising were as follows:

Reynolds ↓ Morris →	Don't advertise	Advertise
Don't advertise	350 350	420 140
Advertise	140 420	190 190

Table 4.4. Profits in millions of dollars at 2021 levels. Values are estimates for similar-sized companies: actual values would vary.

This is a prisoner's dilemma because whatever the other player does, a player is better off advertising, resulting in a significant loss of revenue over the situation where no one advertises. As a result, both companies lost revenue. However, the US government pushed the companies into discontinuing advertising in exchange for immunity from federal lawsuits. This was, from the tobacco companies' viewpoint, a great outcome as it meant that their profits soared, because there was no chance of their opposition reneging and continuing with advertising. Although the government had not intended to increase tobacco company revenues, this was the result of their action.

Vaccination games

The Covid-19 pandemic has provided some significant examples of game theory in action. National governments naturally wanted to ensure the vaccination of their own populations – and some proved a lot better at this than others. Two levels of game played out here.

The larger-scale game was the need to look beyond an individual nation to the world as a whole. Even if the disease was all-but eliminated in one country, international travel and mutations of the virus meant that there was a danger of it being reimported and spreading again in a more virulent form. It was therefore important to ensure that, as much as possible, worldwide vaccination took place. However, there was still a strong argument for the strategy of sorting out your own nation first, if only for the oxygen mask rule.

This is the apparently cruel instruction that airlines give, that in a situation where cabin pressure drops and oxygen masks are deployed, responsible adults should put on their own masks before they help their children or others who depend on them. This is because it is possible otherwise that the responsible adult will lose consciousness before being able to put their own mask on, in which case, the child may not know how to help them. Similarly, even though some individuals complained because their country was focusing on its own citizens first, that country would be less able to help others if it wasn't first stabilised.

However, once your own country is sorted out, even if acting out of pure self-interest in game theoretic mode, it is in that country's interest to ensure as many others in the world are also vaccinated, particularly where there is considerable travel between countries.

Achieving a good balance of the two levels was poorly thought out by the European Union in particular. Rather than simply ensuring its own contracts for obtaining vaccines were effective before helping others, the EU threatened to stop legal exports to other countries from manufacturers in the EU to cover up its own inadequate provision. This could be seen as a clever piece of gaming, but it proved a poor strategy because the vaccine production business is strongly cross-border – if the EU had banned exports, other countries could equally have prevented exports essential to the EU's own vaccine supply chain. At the time of writing, it seems that logic is prevailing – but, for a while, there was a danger of falling into a prisoner's dilemma fail.

At an individual level, such games often achieve cooperation because many human cultures recognise the benefits of cooperation, and reward it in a moral societal structure. Though such moral behaviour can arise from large-scale influences, such as religion, it works best at an individual or small group level where participants are known to each other. Even though, say, two countries may have such a culture in common, it is much harder to implement appropriate cooperative strategies at this level.

Take it or leave it

These classic game theory problems rely on simultaneous decision-making. As we will see in Chapter 5, the best strategies are modified considerably if games are played repeatedly, which is often the case with games that represent social interaction. However, there are other simple games where there is only a single round, but the players take their action sequentially. This wouldn't work with a game like the prisoner's dilemma, where knowledge of the other's player's decision removes the dilemma, but is the essence of a more sophisticated game known as the ultimatum game.

Let's say that there is a potential £10 prize to win. In the ultimatum game, the first player decides how that amount of money is to be divided between the players; after hearing the first player's decision, the second player can either say 'Yes' to accept that split, or 'No' to reject it, in which case the prize is not awarded. The ultimatum game has two types of Nash equilibrium, neither of which is likely to occur in reality. They illustrate the difference between a weak and a strong case. If the first player offers nothing and the second player accepts anything, this is a weak equilibrium, because the second player gets nothing whatever choice is made. The strong equilibrium is for the first player to offer the smallest monetary unit available (in the £10 example, a penny) and the second player to accept anything above zero. In purely financial terms, this is the rational choice – otherwise the second player is literally refusing free money – and it is strong because it is better for the second player (financially) than rejecting the offer.

What's interesting here is that the classic economist's definition of rationality as maximising financial reward fails.

If the first player splits the money evenly – £5 each – then the second player will almost always say 'Yes'. However, if the first player gives themselves a bigger part of the pot, there will be a tipping point at which the second player will say 'No', when they consider their financial gain less important than the ability to punish the other player for selfishness.

For example, with a £10 total prize money, the first player could offer just 10p. This game has been tried many times, and in the vast majority of cases, the second player would say 'No' to this offer. But note again what player 2 is doing – turning down free money. Their decision is not rational by purely economic standards, but it is a natural one to make.

Broadly, some cultures will typically allow the first player to get away with keeping anything less than around 70 per cent of the pot – so in this case giving the other player more than £3 might be accepted – but any less and the other player is liable to say 'No'. In other cultures, something closer to 50:50 is often required. But just as economic benefit is not the only deciding factor, so this outcome itself is not as clear-cut as psychologists tend to present it.

In studies, such games are almost always played with trivial amounts of money. Economics professor Ken Binmore, responsible for some of the game theory auction work we will meet in Chapter 6, has suggested that the effect 'doesn't go away when the stakes are increased' but the reality is that studies which show this have never offered truly life-changing amounts of money.

In a public lecture I give, I ask the audience to take part in an experiment where instead of a £10 pot, we have a £10 million prize. I then ask the audience to stand up and, starting at £1 million, reduce the proportion of the £10 million they would get, asking them to sit down when I get to an amount that they would reject. Not surprisingly, no one has ever rejected £100,000. But bear in mind, this is proportionately exactly the same as the 10p almost universally refused in the low-value version of the game.

People typically start to sit down around the £10,000 mark. Of course, it's one thing to say that you would refuse to be given £10,000, another to do it for real. By the time we get down to £1,000, about half the audience has typically sat down, with a few (arguably rational) people hanging on to around the £10 mark (the equivalent of 0.001p in the £10 version of the game). On one instance only has someone stayed up all the way to £1. As it happens, I knew the person and was able to explain to the audience that he was a Yorkshireman.

This is admittedly a straw poll and a thought experiment where no money changes hands, so can't be taken as definitive evidence. Nonetheless, it is hard to believe that the response is so different from reality that players would reject, say £100,000. Interestingly, although historically the literature has suggested that people's behaviour isn't changed by the size of the stake, more recent research, where genuinely large stakes are involved, has negated this finding, showing that the tipping point for refusing an offer becomes a lower and lower percentage of the total as the size of the stake increases. We're back to the balance of expected value and utility we saw in the Bernoulli game on page 18.

A variant of the ultimatum game, known as the dictator game, takes the second player's contribution away entirely. Here the first player again splits the money, but the second player has not got the option to reject the split. Clearly the purely economic rational decision for the first player is to keep all the cash (the second player has no decision to make), but in practice, depending on social factors, the first player often gives the second player something. It has been suggested that, though less interesting, this game is closer to many real-world situations where an individual decides to give money to someone else while receiving nothing tangible in return.

Pursuing the public good

Another game that comes closer to some real-life situations is known as the public goods game. This reflects situations where there is a widespread benefit from shouldering a personal cost, provided sufficient numbers of people act responsibly. Related activities are vaccination and welfare payments. In effect, this game presents the inverse of the tragedy of the commons.

Here, each player can put as much money as they like into a shared pot. The pot is then multiplied by a number greater than one and less than the number of players and the resultant amount shared equally between the players. Once again, the Nash equilibrium turns out to be in opposition to the best communal outcome.

The Nash equilibrium suggests a player should contribute nothing to the pot. As a free rider, they get the best outcome from the contributions of the rest. However, taking this to the logical conclusion, no one would contribute, so no one would get anything at all. By contrast, if all put their highest possible stake into the pot, the overall benefit is maximised. Note, by the way, that the restriction that the bonus multiplier should be less than the number of players is the deciding factor – if the multiplier is greater than the number of players, the Nash equilibrium is to contribute as much as possible.

Going beyond 2 × 2

Not all games are going to neatly involve two players with two options. If a game still only has two options on one side, but multiple options on the other, we can often eliminate some of the strategies for the player with more than two options if one strategy dominates the other.

Imagine we are back with the caterer we met on page 59. Again, he is dealing with hot and cold weather, but now has a whole range of food combinations to buy, some better in hot weather, some in cold.* Let's focus temporarily on just two of his buying strategies.

Weather: Expected ↓ | Actual →	Hot	Cold
Hot strategy 1	£500	–£50
Hot strategy 2	£400	–£100

Table 4.5. Comparing outcomes of two strategies that assume hot weather.

Whatever the weather, hot strategy 1 is always better than hot strategy 2 – so we can eliminate hot strategy 2 from consideration. In this pairing, hot strategy 1 is known as the dominant strategy. Dominance is one reason for a strategy coming out on top. It's where the Nash equilibrium comes from in the prisoner's dilemma – because giving evidence is dominant for both players.

---

* It may feel as if this game has only one player, but in effect the weather is the other player.

If using dominance to eliminate some strategies fails to reduce a $2 \times n$ game to a 2×2 game which can be solved as before, the game can still be solved by taking a 2×2 part of the game, solving it and seeing if that solution is effective for the other strategies – this involves checking pairs of strategies to find a 2×2 approach that is workable. In practice, game theorists would instead use a graphical solution. These look messy, but are a useful tool.

The approach involves having two axes, one showing the value for one of the strategies on the two-strategy side of the game and the other the value for the other.

Weather: Expected ↓ | Actual →	Hot	Cold
Hot strategy 1	£500	–£50
Hot strategy 2	£400	–£100
Cold strategy 1	–£150	£200
Cold strategy 2	–£200	£300

**Table 4.6. Comparing outcomes of two
hot and two cold strategies.**

If we extend the caterer's options to four strategies, as in the table above, the plot looks like this:

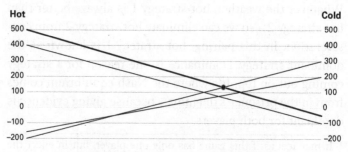

Figure 4.1. Four-strategy plot.

The solution is reached by taking the lines that form the top boundary (marked thickest above) and finding the lowest point on that structure, marked with a dot above. This shows the caterer needs a mixed strategy, based on a combination of hot strategy 1 and cold strategy 2, with values calculated in the usual 2 × 2 approach.

In a game with more than two columns but only two rows, the graphical solution involves finding the bottom boundary and identifying the highest point on that.

Rock paper scissors

Of course, real world two-player games aren't always limited to 2 × n choices, but rather can be m × n with a range of strategies available to each player. Perhaps the simplest is the familiar mechanism for selecting between two individuals, rock paper scissors, a 3 × 3 game giving each player the same three choices.

Player 1 ↓	Player 2 →	Rock	Paper	Scissors
Rock	0	–1	1
Paper	1	0	–1
Scissors	–1	1	0

Table 4.7. Outcomes for rock paper scissors.

Here we are back to a zero-sum game like noughts and crosses, though there is no possible saddle point because all the column maxima are 1 while all the row minima are –1. We are unable to eliminate any strategies, as none is dominant. There is no Nash equilibrium for a single game of rock paper scissors, because there is no combination that is the best for each player given the choice of the other.

Because of the symmetry in this game, there is no need to carry out the calculations necessary for mixed strategies in some of the games above. Here the Nash equilibrium solution is for each player to choose between the three options with equal chances, playing each ⅓ of the time. As always, should one player not use the optimum strategy – say, always playing rock – the other player should then move to a strategy reflecting that error, in this case always playing paper.

For other multiple-strategy $m \times n$ games, we can apply, as before, checks for a saddle point and for dominance. However, if the structure does not resolve down to a simply solvable 2×2 game, while a solution will remain, finding it becomes more and more complex. A graphical approach can still be used, but we are into multi-dimensional space that is not easily drawn and has to be solved with a computer. The mathematics becomes too complex to work this out easily in a straightforward demonstration (typically in the so-called pivot method, which involves multiple manipulations of the table) but is always theoretically soluble.

Anticipating the opponent

In games (whatever they represent in the real world, whether it's a board game, economics or human behaviour in general), we rarely have complete knowledge of the other players' strategies. As a result, there is always a danger of recursive arguments, where a player can assume what other players' strategies are – but if another player knows what the first player is doing, they can act to counter this.

This can be seen in a simple form when playing rock paper scissors. There is some evidence that men are relatively likely to choose rock as their first option. This means that if player 2 knows this and is playing against a man, it would be to her benefit to choose paper. However, if player 1 were also aware of this tendency, and thought that player 2 would go for paper, player 1 should actually choose scissors. But if player 2 suspected this, player 2 should choose a rock ... and so on.

Similarly, if I am playing, and I am aware that my opponent knows that the best strategy long term is to choose at random, then I might calculate that there's a better chance that their second choice will be different from their first, as there are two different options, but only a single unchanged option. So, if we both initially went for paper, I might reason that my opponent's next choice would be more likely to be rock or scissors ... unless the other player was also trying to second-guess my strategy.

The risk of a player familiar with game theory thinking through strategy in this way, though, is that most people playing rock paper scissors don't adopt such a subtle approach. Last time I played someone, they started with paper. I avoided rock, because I wondered if they would assume a start of rock on my part, and probably should have gone for scissors, but without any good reason went for paper as well. However, when afterwards I asked the other person why she had chosen paper, she said it was because it was her favourite of the three because of its neat rectangular form.

We always have to bear in mind with game theory that it assumes rational behaviour on the part of the players, with the proviso that 'rationality' may well be based on something more than pure economics. Unfortunately, in the real world, we often make small decisions for irrational reasons, because they don't really matter much. Most of us accept life is too short to start weighing up the strategic pros and cons of, say, which chocolate to choose from a box. We hope that we are being rational in important decisions – though arguably even here it may not be the case.

I guess you're right

Sometimes the structure of the game can be based on our beliefs of what choices others will make. A simple game to illustrate this is the guessing game (sometimes known as a beauty contest, the name given to a concept devised by English economist John Maynard Keynes). So, for example, we could ask a group of people each to choose a number between one and ten inclusive. The person who guesses closest to two-thirds of the average of all the values chosen wins the prize. There is no point choosing a number bigger than seven, because even if everyone chose ten, the value people were aiming for would be 6.666... – to which the closest value is seven. In the terms we've already used, seven is a dominant strategy over all guesses larger than seven.

However, if everyone playing were knowledgeable enough to work this out, we are really only asking players to choose values between one and seven. If that's the case, then two-thirds of the average is not going to be more than five. Which suggests that there is no point going for a number larger than five. But if everyone is only choosing numbers between one and five ... and so the process continues. In the end, everyone using this logic could only ever choose 1

– the Nash equilibrium – and everyone would win (though should there be a single prize to be split, this may result in a poor reward).

In the real world, though, people asked to take part in this game may not think things through in such detail. Let's say everyone chose totally randomly. What would be the most likely outcome? The number five often springs to mind as being around the middle, two-thirds of which is 3.333…, to which the closest value is three. However, in reality, the average of the numbers one to ten is 5.5, two-thirds of which is 3.666… – producing a winning value of four. Out of interest, I ran a couple of 1,000-person games using random selection of a whole number between one and ten: two-thirds of the average came out as 3.736 in one case and 3.754 in the other.

Things could be different in the real-world game, though, because it is often said that seven is the most likely number for people to choose when asked to pick a number between one and ten. That being the case, it is interesting to consider what would happen if seven were more likely to be chosen than the other options. Even if half of the people chose seven, the winning number would still be four, but if *everyone* chose seven, the winning number becomes five.

A variant of this game was tested out for real, twice, in the *Financial Times* newspaper, in 1997 and 2015, in a competition devised by the American economist Richard Thaler. For Thaler's version of the game, the range was 0 to 100, and participants were asked to make their guess 'as close as possible to two-thirds of the average guess'. In the 1997 trial, the prize was two business-class plane tickets from the UK to America, while in the 2015 version (run by English economist and writer Tim Harford on Thaler's behalf), the prize was 'a posh travel bag' and a copy of Thaler's latest book. (Interestingly, 1,382 played for the tickets, but only 583 for the travel bag.)

The readership of the *Financial Times* is likely to be more numerate than the population at large: though there was a scattering of entries across the range in the latest poll, the most favoured guess was one, followed by zero, then 22. There was also a small peak at 42, presumably from fans of Douglas Adams, whose book *The Hitchhiker's Guide to the Galaxy* gave 42 as the answer to the ultimate question of life, the universe and everything. There was also a peak at 100 from a coordinated group of entrants who admitted that they were intentionally trying to force up the average to make rational decision-makers less likely to win.

Because of the distribution of solutions, the winning guess in 2015 was twelve, while in 1997, the distribution was similar with a win for number thirteen. In the end, there were around twenty correct entries of the 2015 game – the winner was chosen for giving the best explanation of their choice.

Like the other games we have explored so far, Thaler's was only played once – but the real world often involves repeated interactions, a reality that game theorists were soon to acknowledge.

IF AT FIRST YOU DON'T SUCCEED

5

Groundhog Day

We have already seen the impact on game theory of mixed strategies, when theoretically playing a game more than once is used to produce Nash equilibria to support a single decision. However, there is a more profound consequence when repetition is for real, and particularly when the potential for future repetition of the game is open-ended, that is, in a situation where neither party knows if or when the sequence of games will finish. This might seem odd to contemplate, because a sequence of games is usually associated with a fixed-length tournament, whether we are thinking about rock paper scissors, chess, or tennis. However, bear in mind that game theory has a wider remit than literal games, and many human situations have no predictable ending to a series of interactions.

It is this possibility of open-ended repetition that game theory uses to explain reciprocity – the idea that one person is willing to accept a reduced outcome for themselves, benefiting someone else, because at some point in the future the position is likely to be reversed. This means that, taking the long view, it is a better strategy to think of others, rather than adopting the economist's starting point for rationality and acting purely selfishly.

Reciprocity does not, however, extend to cover altruism, where an individual decides to take a reduced outcome for themselves to provide overall benefit in a single game or without any potential future gain. Here, simplistic game theory based solely on financial gain is not sufficient and we need to bring in both the positive feeling generated by doing something for someone else and the cultural and social framework of caring for others.

Zero-sum thinking

A series of experiments in the late 1950s and early 1960s at Ohio State University pitted students against each other in a sequence of repeated simple two-player games. What came across strongly here is the unsurprising result that people are not good at judging the implications of numerical outcomes pertaining to themselves. But perhaps more interesting was the fact that the players appeared to be applying the wrong model to the games.

In the Ohio State games, students had a red and a black button to choose between and were provided with a pay-off table. One of the games had a table as shown in Table 5.1.

Note how bad playing red is here. It is the antithesis of a Nash equilibrium. Whatever the other player does, a player is worse off if they push the red button. Yet in the study, the students hit red 47 per cent of the time.

Player 1 ↓	Player 2 →	Red	Black
Red		0 0	1 3
Black		3 1	4 4

**Table 5.1. Ohio State game outcomes
where red is always worse.**

Before considering why such an illogical strategy was
deployed, we have to insert a proviso. The results of many
social science experiments have been called into question
over recent years because the sample sizes were too small*
or the experiments were not rigorously designed, meaning that they did not actually test for the effect supposedly
being measured. All too often, these studies have also either
employed cherry-picking – where only results that are sup-
portive of an argument are recorded – or have been guilty
of so-called p-hacking, where researchers look at the data
produced in a wide range of ways until they find a particular
combination of factors where there is statistical significance,**
even though this will happen purely coincidentally if there
are enough ways to arrange the data.

Another issue, as was the case in the Ohio experiments,
is that social science experiments often use students as test

---

* Sample size refers here to the number of people taking part in
the experiment. To have a meaningful representation of the public
at large often requires many hundreds or even thousands of participants. Some of these studies involved fewer than ten people.
** Statistical significance is a measure of how likely it was that the
effect observed could have happened purely by chance. Just because
a result is significant in this sense does not mean it is important. A
result could be totally trivial, without any meaningful implication,
but still statistically significant.

subjects, individuals who are extremely unrepresentative of the population at large, in a wide range of factors from age and education level to ethnicity and degree of deprivation. The many problems in social science studies have led to the so-called replication crisis. When many older studies were repeated in a more careful fashion, they failed to produce the original results. In 2015, an analysis of 100 psychology studies found that, on average, effects were half the size originally noted and that where 97 per cent of the original studies found significant effects, only 36 per cent of the attempts to reproduce the studies did. We have to take the results of all older psychology and sociology studies with a large pinch of salt. But even if the 47 per cent value in the Ohio State study is an exaggeration, it seems bizarre that anyone would play red.

It has been suggested that the irrational choice of red was because we are used to games being zero sum. Certainly, many board and card games do have a zero-sum structure. And this makes it easy for anything that feels like a game to be treated as if the aim is to beat the other person, not to devise the best strategy, which may include the possibility of collaboration. The suspicion was that in the Ohio State games, getting the same outcome as the other player was considered a draw rather than providing a win or loss. If one player chose black, the other felt that choosing black as well each time was a poor strategy, even though the table makes it clear that it is not. Although a later series of games was designed to avoid gameplay terminology – for example, referring to the interaction as a transaction rather than a game – the setup of such an experiment feels more like a win-or-lose game than a real-world interaction with another human and may have influenced the decision-making process.

Grim and punishment

Repeated games have been explored in a number of tournaments where a range of computer algorithms go up against each other to see which strategies come out on top over time. Most frequently explored in this fashion is the prisoner's dilemma. Many of the extended play strategies work on a degree of retaliation in an attempt to force the opponent into cooperation. The most vigorous of these, known as 'grim' involves being cooperative (not giving evidence) until such a time as the opponent reneges (gives evidence). From then on, and forever, the strategy says not to cooperate. One strike and you're out.

Grim is an all-or-nothing strategy. If a human player is aware that the other will definitely undertake this strategy, the rational thing to do is to cooperate. However, since in general we don't know the other person's strategy, many strategies will involve reneging occasionally to test the water; grim's all-or-nothing approach means this will have a negative effect on the overall outcome for someone trying their luck. In practice, there is a better winning strategy for repeated prisoner's dilemma games called tit for tat.

At its simplest, this strategy involves starting off by cooperating, and from then on playing whatever move the opponent played in the previous game. The idea feels like a classic 'eye for an eye' human response. However, to be an effective strategy it needs some subtlety. If we think of automated algorithms playing this exact strategy, it becomes an oscillating equivalent to grim. Let's look at the outcome table used before and see the results. This is the 'positive' version of prisoner's dilemma where the numbers are the amount won, so that we can see accumulated winnings over repeated play – the jail time version rapidly becomes confusing. With the numbers used above, the outcome table looks like this:

Player 1 ↓ | Player 2 →	Cooperate	Renege
Cooperate	£5 £5	£6 0
Renege	0 £6	£1 £1

**Table 5.2. Outcomes for positive
version of prisoner's dilemma.**

Let's imagine looking at the outcome of the first six games. When both players use a grim strategy, resulting in complete cooperation, the outcome would be to win £30 each. If the players both use the most basic form of tit for tat, then the outcome would be the same, as the rule says to cooperate unless the other player gives evidence of reneging. But let's say one player chances their arm and reneges, after which both play tit for tat.

We would have a sequence like this:

Player 1		Player 2	
Renege	£6	Cooperate	0
Cooperate	0	Renege	£6
Renege	£6	Cooperate	0
Cooperate	0	Renege	£6
Renege	£6	Cooperate	0
Cooperate	0	Renege	£6

**Table 5.3. Outcome sequence in first six games with
tit for tat strategy where one player reneges once.**

The outcome is only to win £18 each – significantly worse than grim versus grim. The players are locked into an unbreakable pattern of failure and retribution. Tit for tat also does not produce the best result for player 1 if player 2 is playing the (admittedly unlikely) strategy of always cooperating, even if player 1 reneges. (Short of playing against a saint, this only tends to occur when testing computers that cannot change their strategies.) If the tit for tat strategy is used against always cooperate there would be cooperation throughout, resulting in winnings of £30 each. The most rewarding strategy against someone who always cooperates is to always renege. Here, every time the player who doesn't cooperate scores £6, winning a total of £36 overall.

Where tit for tat comes into its own is where an opponent is playing a more reactive strategy. Here, after chancing their arm, the other player realises they will suffer if they continue to renege and so begins to cooperate, which is rewarded by the tit for tat player, producing a sequence like this:

Player 1		Player 2	
Renege	£6	Cooperate	0
Renege	£1	Renege	£1
Cooperate	0	Renege	£6
Cooperate	£5	Cooperate	£5
Cooperate	£5	Cooperate	£5
Cooperate	£5	Cooperate	£5

**Table 5.4. Outcome sequence if player 1
reneges but then cooperates.**

After player 1 reneges, player 2 responds with tit for tat. Provided player 1 realises he or she is in trouble and is prepared to cooperate twice in a row to compensate, the repeated game comes back to the stable state. Over the first four games it has the same outcome as a give/take oscillation, but after that, the strategy starts to pull ahead, so that by the time the players reach six games, each has won £22 because the average each player gets from cooperation is £5, while the average in an oscillation between cooperating and reneging is only £3.

If it's possible that the other player is playing 'always cooperate' – indistinguishable from tit for tat until there is a challenge – it is arguably worth taking the player 1 strategy above. In the example shown, player 2 is using tit for tat. If, instead, they were always cooperating, then their response to the second game would also be to cooperate – in which case, player 1 should stick for the moment to reneging.

Unnatural selection

The way that modern artificial intelligence systems use the process of machine learning to get better at playing a game is to evolve their strategy. When political scientist Robert Axelrod ran a number of computer-based trials of different strategies for repeated prisoner's dilemma at the University of Michigan in 1980, one approach taken was to develop a kind of evolutionary system. He and his team ran a whole range of strategies against each other in a tournament of repeated games and saw how they scored.

The strategies were then allowed to undergo competitive selection, in that the number of instances of a particular strategy in each subsequent tournament was set by the scores that the strategy got in a previous tournament. Initially, a number of strategies survived, and a number died out. But some surviving strategies depended for their success on having particular opponents available. In effect, the surviving strategies were like predators, some of which were fussy about which prey they took on, others of which could take on a whole range of opponents well.

As a result, after a while, some of these remaining strategies also died out because their prey was no longer available. In many circumstances, tit for tat remained the superior approach. The only circumstance in which it had the potential to die out was in an environment with a high concentration of players who constantly reneged.

Things get a little more complicated here, depending on the tournament's structure. There are three possible ways for the gameplay to proceed. Players could always play against the same opponent; they could play against other opponents, selecting a new player every time but continuing their strategy; or they could select a new opponent every time, keeping track of their strategy against that player.

In the first style, whichever strategy has larger numbers, whether using tit for tat or always renege, is likely to dominate, because if a tit for tat plays another tit for tat, it will always do better than a reneger, but if it plays a reneger from the first game it will do slightly worse, as the reneger wins the first game and thereafter they both draw. Similarly, in the second style of play, any tit for tat player that has played a reneger will never be able to recover after that point, so it depends how many games they will typically have played before being matched against a player that reneges.

In the third style, however, even a small number of tit for tat players will gradually take over, as every round they play against another tit for tat will give them a high score because their 'trust' of the other player is retained. A player who always reneges will only be trusted once, so will gain only a single large score from each tit for tat player, but every time a tit for tat player plays another tit for tat they will both benefit.

Starting at the end

The success of tit for tat suggests that this is the obvious way to play in a repeated prisoner's dilemma game. And for an unending game (or one where we don't know the ending, such as a lifetime) this generally applies. However, if the players do know how long the game is to be run, the strategy can seem to be undermined.

Let's imagine a repeated prisoner's dilemma with the same outcome table as above, but where the players know that there will be exactly ten games. The last game becomes a special case. There is no opportunity for a tit-for-tat response to a player's actions. Therefore, it can be tempting to renege as the last choice. But look what this does. Now the penultimate choice is a special one if the last choice is already fixed. So yet again it would seem that on the ninth game a rational player should renege and maximise their return.

This 'backward induction' mechanism taken to the extreme suggest that players should renege every time, taking us back to the bad old days of the single-play prisoner's dilemma. Luckily for rational players, though, there is a significant issue with this argument.

If both players are aware that they will each reason in this fashion, then the outcome becomes inverted. Both should opt for cooperation in the final game to avoid the non-Pareto-efficient outcome – and this ripples back through the games. Such reasoning, however, doesn't prevent each player from thinking that if their opponent adopted this reasoning, then they should defect on the final game anyway, as they will win the most that way. There is a danger here of the unescapable circular logic that emerges from 'if I think that you think that I think that you think…'

In practice, it is arguable that players would decide that because of the strong risk of getting locked into a renege/renege scenario, logical players would in practice both cooperate each time, except possibly taking the risk of reneging on the last round.

Pumping irony

A real-world process that plays out as a repeated game is a price war. Every now and then, in a situation where there are a few dominant players supplying a commodity market – one in which we don't really care which brand we buy – prices will suddenly plunge. But there is more happening here than simple undercutting. This is perhaps most obvious with petrol (gasoline) pricing.

In effect, what happens to the prices at the pump is a repeated prisoner's dilemma game. Here, the equivalent of reneging is cutting your price to below that of your competitors, while keeping roughly the same price is cooperation. When one competitor is noticeably lower priced than the rest, their sales will shoot up as word gets around. If the competitors also renege – which, remember, is the economically rational thing to do – the fuel price will go through a series of drops until it settles again.

This should be resolved at just above the cost of providing the fuel to the filling station operator. In practice, though, price-cutting can go further, particularly as the cheapest petrol is often found at supermarkets, which may be prepared to make fuel a loss leader on the assumption that some fuel customers will also use the store for other purchases.

The interesting thing, though, is that most of the time the prices at equivalent outlets are very similar. (There is always significant variation between, say, a filling station on a motorway and at a supermarket, but comparable sites for different fuel vendors are usually similarly priced.) It might be thought that this is the result of illegal price-fixing collusion, but collusion isn't necessary. The companies are aware that the only way to maintain a reasonable profit is to charge a similar amount to their competitors, so the tendency is to observe competitors and home in on a similar pricing strategy.

This is a repeated prisoner's dilemma scenario where the players see the overall advantage of cooperation and the dangers of tit-for-tat responses, so (usually) cooperate without the need to communicate.

I'll show you mine if you show me yours

Up to this point, we have mostly looked at games where the players don't know what the other player has chosen before making their decision. In a single game this can make for some unpleasant rational choices but, in repeated games, behaviour in one round can be rewarded or punished in the next round. In many actual games, though, the information that comes from repetition originates not from repeated simultaneous play, but from players taking their turn openly, one after the other. Here the games take on a new structure.

Take the simple game mentioned on page 87, sometimes called the battle of the sexes, where our outcome table looked like this:

Player 1 ↓ | Player 2 →	Movie	Meal
Movie	2 3	0 0
Meal	0 0	3 2

**Table 5.5. Outcomes for the battle
of the sexes game.**

To make this game work, we had to come up with an unlikely scenario where the two players could not communicate and did not know what the other player would do. But what if the game worked more like a board game where players take turns openly? For example, player 1 could have left work first, travelled to the cinema and then messaged her location to player 2. Here the kind of tree we saw on page 51 can be useful, but in this particular example, there are no probabilities involved.

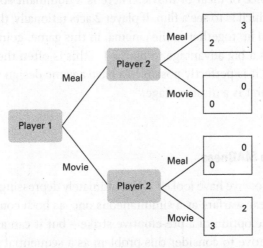

Figure 5.1. Battle of the sexes game tree.

Unlike the simultaneous game, player 1 is in the position to make her decision first, and she will opt for a movie. Now we can prune off the part of the tree involving the other decision, as this will never arise. So, player 2 is presented with a simpler decision tree for what is known as the subgame:

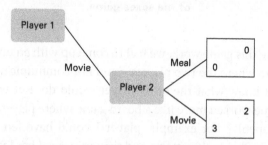

Figure 5.2. Pruned battle of the sexes game tree.

Now there is only one decision to be made – player 2 has the choice of meal or movie. There is a dominant solution here, which is to see a film. If player 2 acts rationally, the pair will end up together at the cinema. In this game, going first delivers a big advantage to player 1 – this is often the case, though it is perfectly possible to have a game design where going first is a disadvantage.

Back to MADness

Up to now we have looked at the ultimately depressing game of nuclear warfare as a simultaneous one, as both countries have the option of a pre-emptive strike – but it can also be informative to consider this problem as a sequential game. Before we look at the detail of mutually assured destruction, let's spend a moment longer on the cinema/restaurant decision above.

What if player 2 said *before play*, 'Whatever happens, I am going to go for a meal. It's up to you if you want to join me'? Player 1 then has to decide if this is a credible threat, or just posturing. Logically speaking, in this case the threat is not credible – because according to the game's outcomes, player 2 would rather go to the cinema with player 1 (reward of 2) than go for a meal on his own (reward of 0). If player 2 behaves rationally, he will see the film despite his threat – but the expression 'cut off your nose to spite your face' was not coined for a situation that never occurs.

Individuals can be stubborn and take an action that results in a non-optimal outcome to save face or to maintain a power relationship. Children can sometimes stick to a bad choice from a lack of ability to work through the logic. Adults can also be stubborn in a self-damaging way, but sometimes requirements like saving face are genuinely more important if this is required to maintain a position in an ongoing series of games. Think, for example, of that game that is familiar to any parent – the coercive threat.

Imagine that a parent wants their child to take a particular action – or to stop doing something. So, in the past I might have asked one of my daughters (both now grown up) to clear up her bedroom. Let's follow through the sequential game. She responds with 'I'll do it later'. However, I know perfectly well that this means it will never happen. So, I come back with 'No, do it now', only to get a straight 'No!' in response. By now I am angry. Under the pressure of the moment, I make an apparently non-credible threat. I tell my daughter that if she doesn't do this, she won't be allowed to go to her friend's birthday party tomorrow.

This should be a non-credible threat, because I know that if I prevent her going, I will be sad that she is missing out (the friend is well known for having the best birthday party of the year), and we will get a negative response from the friend's parents, who are expecting our daughter to go. Yet if I back down, in the future my threats will not be taken seriously (and I will be told off in no uncertain terms by my daughter's twin sister for not following through). Even though, when looking at the outcome of the single game we are currently playing, I should *not* carry out this threat, because it is lose-lose for my daughter and for me, I *should* carry out the threat because this is not a one-off game and the consequences for future gameplay will be strongly negative. (Of course, I should not have made the threat in the first place, but once it is made, the game is in play.)

Scarily, this apparently trivial domestic punishment game is one that has direct parallels for the nuclear war scenario and for mutually assured destruction. The MAD strategy only works if massive retaliation is a credible threat. In the run-up to the 2019 UK general election, Jeremy Corbyn, then leader of the Labour Party, said that he would never push the nuclear button. Had Corbyn won the election, Britain's nuclear deterrent would have become a non-credible threat. Then, if a potential attacker felt that making a nuclear attack on the UK was preferable to not doing so, it would only be logical to carry out that attack.

Of course, we can't know that any country would adopt an offensive nuclear strategy. An enemy country may, for example, have moral objections to carrying out a nuclear attack, meaning that it was not their preferred strategy, even if the possibility of retaliation had been pruned from the game tree. However, it is hard to see that making the

UK's threat non-credible would be beneficial, and it is likely that had Corbyn been elected there would have been some attempt to rebuild the credibility of the threat, perhaps by modifying the mechanism for authorising a nuclear strike.

The game theory bible

There are a handful of lengthy science and maths books in history that can be seen as forming distinct milestones in progress. Often, they are far better known than they are read (in part because they are mostly distinctly indigestible). Examples that spring to mind are Roger Bacon's *Opus Majus*, a remarkable exploration of thirteenth-century science; Isaac Newton's masterpiece *Philosophiae Naturalis Principia Mathematica*; introducing his laws of motion and gravity; and Alfred North Whitehead's and Bertrand Russell's *Principia Mathematica*, named with a nod to Newton, that pulled together the foundations of mathematics. Game theory has its equivalent in John von Neumann and Oskar Morgenstern's monster tome of over 600 pages, *Theory of Games and Economic Behavior*.

Although much of the book consists of dense mathematical equations, it also contains material that gives a good feel for the mindset of those who developed game theory. Morgenstern was an economist, and the aim of the book was purportedly to bring the power of game theory to bear on economics; but in practice it contains plenty of material on pure game theory. What's more, *Theory of Games* has never been popular with economists, in part because the book is distinctly critical of the way that maths had been used in their field.

The authors note: 'Mathematics has actually been used in economic theory, perhaps even in an exaggerated manner. In any case its use has not been highly successful.' Most scientists and mathematicians would have agreed with this analysis, but it was hardly likely to endear the book to economists. The authors go on to point out that most sciences depend on maths and explain why they feel economics struggles with it. 'The arguments often heard that because of the human element, of the psychological factors etc., or because there is – allegedly – no measurement of important factors, mathematics will find no application, can all be dismissed as utterly mistaken.'

Von Neumann and Morgenstern go on convincingly to pull apart these arguments. They point out that the real issues are that economic problems tend not to be formulated clearly, but in vague terms, and that mathematical tools had rarely been used appropriately in the field. Although maths is now a much more significant part of economics than it was in 1953 when the book was written, it could be argued that the latter is still true (see David Orrell's book *Economyths*, listed in the further reading section at the back of this book, for some of the misuses of mathematics in economics).

However, von Neumann and Morgenstern were incorrect in one regard: although game theory has proved immensely valuable in better understanding decision-making processes, it has failed to provide a mathematical mechanism to help economics become a more legitimate science. Game theory is extremely useful from the psychological and practical viewpoint of understanding decision-making and human thought processes – but it is not a practical economic tool in most circumstances.

Newcomb mixes it up

One scenario that particularly underlines the way that game theory throws a light on understanding a process, without providing a decision-making tool, is a situation where opposing views of the decision structure produce conflicting outcomes. The best-known example of such a problem, devised by American physicist William Newcomb in 1960, involves a game that can easily be envisaged as a part of a real TV game show.

Rather like the show *Deal or No Deal*, in this game the contestant has to choose between boxes containing different amounts of money without knowing their contents. Here, though, there are just two boxes. The first contains £1,000, which is known by the contestant. The second either has £1 million inside, or nothing at all. The contestant can either choose to open both boxes, or just to open the second box – and she wins whatever is in the box or boxes she opens.

This would seem like a totally trivial choice, as it seems inevitable that opening both will maximise value, yet the game has an odd twist. The team behind the game show decide what will go in the second box based on an in-depth study of the contestant. They have everything available, from her medical and educational records to interviews with her close friends and family. From this research, the team makes a choice. If they think that the player will choose only the second box, they put £1 million in it. If they think that the player will choose both of the boxes, they put nothing in the second box.

To help the player make their choice, a scoreboard at the back of the studio keeps a tally of how many times the team has guessed right during the long-running show, and how many times they have guessed wrong. When you are lucky enough to be selected to play, there have already been 1,000 games played, and the team has only made the wrong guess once. This means that 99.9 per cent of the time the team have guessed correctly what the contestant would do.

Would you choose to open box 2, or both boxes?

The fascinating thing about this game is that the majority of potential players know what they would do – but the decision is split roughly 50:50. And in game theory terms this makes sense because there are two ways to analyse the outcome, one supporting opening box 2, and one that favours opening both boxes.

The first strategy is based on expected return. Assuming that the team continue to perform at the same level they have for the last 1,000 games, the expected return from picking box 2 only is $1,000,000 \times 0.999 + 0 \times 0.001 = £999,000$. The expected return from picking both boxes is $£1,000 \times 0.999 + 1,001,000 \times 0.001 = £1,999.10$. It's a strong argument in favour of picking just box 2.

But let's look at the game in a different way.

Player ↓	Team →	Both	Box 2	Row minima
Both	£1,000	£1,001,000	**£1,000**
Box 2	0	£1,000,000	0
Column maxima	**£1,000**	£1,001,000	

Table 5.6. Outcomes from Newcomb's game combining player's choice and team prediction.

If we consider a single game, then there is a saddle point, telling us that the rational strategy is for the player to choose both boxes. Whichever strategy the team has chosen, the player will be £1,000 better off by choosing 'both' – it's the dominant strategy. This is reminiscent of the prisoner's dilemma. There is a clear strategy to be taken, even though taking the alternative strategy of trusting the team to predict the player's choice correctly is likely to result in a better outcome.

There is no 'right answer' here. We're back to having to rely on utility. If £1,000 would be transformative, so the player could not afford to lose it, she should go with the saddle point minimax solution and open both boxes, because that £1,000 would be guaranteed. But if the player could survive not winning £1,000, she would be best to go with the expected return and open box 2, with an excellent chance of winning £1,000,000.

Context matters

Newcomb's game may feel artificial, but there is a well-established and frequently experienced psychological effect that could have an influence on the decision made by the player of a Newcomb game. This is that the context in which an amount of money is placed influences how significant we consider that amount to be. We can lose a grip on the utility of an amount when it is put in the context of a significantly larger figure.

The guaranteed amount in the Newcomb game, £1,000, is a significant figure for the majority of people. Most of us, for example, would be devastated to lose a wallet containing £1,000. Similarly, if it were possible to buy the same holiday from one provider for £2,000 or from another, with a bit more effort, for £1,000, the majority of people would consider it well worth taking the trouble to save £1,000.

However, in a house purchase for, say, £300,000, most people will treat a £1,000 difference as trivial and not worth putting in much effort for. It is exactly the same gain or loss. The utility is exactly the same. But because of this psychological context effect, the value seems less. Taking a game theory approach, we should always be careful to consider the utility of an amount in isolation.

Do you know what they know?

In the real world, we rarely have as perfect a set of information as do players of a game show. On the programme, the rules are clearly stated. We don't necessarily know the rules of life 'games', we don't necessarily know what the other players' strategies are, and we may not have clear information on the outcomes of adopting a particular strategy. Some games are particularly biased because one player has information that the other doesn't – so-called asymmetric information – while in others, the information may not be known by anyone. A player could even be deliberately fed incorrect information.

Often, in a commercial transaction there is asymmetric information. In a second-hand car sale, for example, the seller knows significantly more about the vehicle than the buyer does. In such circumstances, the buyer might sensibly either be prepared to pay someone else with expertise to check the condition of the car, or pay a premium to buy a second-hand car from a reputable dealer who will fix any problems that arise. Although they may not recognise it as such, the point of the buyer paying extra money is to increase the information that they have available to them, balancing the game.

Equally, such an asymmetry can be something that the seller of a product needs to overcome. A lack of information can make a buyer reluctant to purchase an unfamiliar product, making it difficult to bring a new item to market. It is not enough for the seller to give the buyer information, as data from such a potentially biased source is unlikely to be trusted. Instead, mechanisms are employed to provide acceptable information, including free samples and reviews from trusted experts or known sources. More recently, thanks to the internet, mass reviews and ratings from other customers have also been a mechanism for increasing buyer information, though in some circumstances this approach can lack buyer trust, as sellers have found ways to game the system, loading the reviews in their favour.

Games red in tooth and claw

Game theory was originally designed to explore the nature of human behaviour, but there is no reason why it shouldn't be extended to take in other species, and from the 1980s onwards there has been an interest in applying game theory to the wider world and particularly to the evolution of behaviours.

At the heart of our understanding of biology is the theory of natural selection or 'survival of the fittest'. It should be stressed that this does not mean 'survival of the best' on any absolute measure of what 'best' means. It is rather that at a particular time in a particular environment, the species or mutant variant that was best suited to survive would be more likely to live long enough to reproduce and pass on its genes to its offspring, resulting in that species or variant thriving for the moment. Given the way that the rational outcome of game theory can often run counter to cooperation, and the idea of natural selection also seems to run counter to cooperation, it might seem inevitable that nature always goes for the quick reward. And often this is true – but certainly not always.

So, for example, some predators allow potential prey species to come near to them without attacking them, if the predators can benefit from cooperation. Think of the birds that eat ticks and other parasites off carnivores, sometimes even straying into the predators' mouths to pick out unwanted guests. While both the bird and the carnivore potentially benefit from cooperation, it might seem that the pure game theory approach suggests the predator should renege on the unwritten deal and snap up the bird.

We can't argue that in the case of the predator this is likely to be the result of the animal thinking through the long-term consequences of its action, working out how such a strategy would inevitably mean losing the benefits of the cooperation. This seems beyond the capabilities of most species. So, what is happening here? One possible solution is that there is a difference between the kind of reasoning a human might employ and learned behaviour, which is common to most animals.

While we have little evidence for non-human animals thinking about long-term futures, it is entirely possible for even those with very limited brains to learn that a behaviour will reward them and to subsequently repeat that behaviour. Much work has been done, for example, showing this with pigeons – not exactly the brightest birds in the box. Anyone who has had a goldfish pond will know that feeding fish at a regular time from the same spot will result in them heading for that part of the pond at around the right time.*

Rather than demonstrating logical capacity, learned behaviour reflects the way that brains are self-organising systems, reinforcing links between neurons that are used regularly. If a particular response has a positive outcome, it is likely to be repeated – and if that continues to be the case, the probability of making that response increases. Of course, in any specific cooperative prey–predator relationship it is entirely possible things will go wrong and a quick snap of the jaws will end the cooperation. But if, more often than not, the prey survives the interaction and things go well for both animal players, it is entirely possible over time for such a relationship to build up.

Although the behaviour itself will not be passed on to the animals' offspring, predator and prey that are more genetically inclined to take the cooperative stance in such a circumstance may be more likely to reproduce and pass on that genetic inclination, while the young of the predator species may then learn the behaviour before they are too much of a threat to the cooperative prey animal. Similarly, there is often cooperation within species, where the balance is more equal than in a predator–prey situation. Some analysis has been done that suggests such cooperation has a basis similar to the tit for tat strategy in the repeated prisoner's

---

* The goldfishes' learned behaviour is a clear demonstration of the inaccuracy of the fallacy of their three-second memory.

dilemma – perhaps not surprising considering the strategy's evolutionary survival in computer simulations.

The existence of cooperative behaviour, of course, is as true with humans as with any other animals, but there are many activities that are special to human interaction, and one of these has come to the fore in the modern use of game theory.

GOING ONCE, GOING TWICE 6

Bidding for victory

The most important single practical use of game theory has arguably been in the design of specialist auctions. As we saw in the opening chapter, these have been used highly effectively in the distribution of bandwidth for mobile phones.

Traditional auctions are very simple games. The most familiar type, known as a forward or English auction, simply sets a minimum value for an object being sold (its reserve price) and as long as someone bids at least this value, the highest bidder purchases the item for the amount that was bid. There are few strategic choices available, other than setting a ceiling for your bids and sticking to it.

Online auctions, such as eBay, use a slight modification of the traditional auction where the winning value is determined by the size of the second-highest bid. Here, bidders set their ceilings in advance, and the system advances towards these ceilings, depending on the amount offered by other bidders. If the opposing highest bid for an item is £50 and you win the item after bidding £70, you do not pay £70 in such a system but either £50 or £50 plus a small increment, the point being that you were prepared to pay more than the second highest bidder, so deserve to win.

A third common auction format is the Dutch auction. Here, the selling price starts at a high value and gradually falls. When a potential purchaser feels they would be happy to pay the amount the auction value has fallen to, they bid and immediately win.

Unlike the two-player games we have mostly encountered, auctions are games that take place between the seller, represented by the auctioneer, and any number of buyers. One powerful strategy that buyers could use is collaboration. So, for example, buyers could discuss among themselves which items they each want and agree not to bid against each other on those items, minimising the price paid. Such a bidding 'ring' is illegal in most countries, but it can be difficult to prove that the strategy is being used, and the approach has certainly been widely employed.

Game theory is also relevant to interactions between buyers where there is no cheating involved. A Dutch auction, for example, is an inverted version the game of chicken, which, as we have seen (page 88), is often described as two cars driving towards each other down the middle of a road. Each player is daring the other to go further, but eventually someone gives way. In chicken, the person who sticks in the game longest wins. The inversion comes from the fact that in the Dutch auction, the person giving way first is the winner (but pays a higher price the sooner they 'swerve'). As in other games, if a player can gain knowledge of another player's strategy, then they are in a much better position to optimise the outcome from their viewpoint.

When game theory started to be applied to the design of auctions, the aim was not to make things easier for the buyers, but rather to push buyers towards paying more for their purchases. Some might consider this an unethical example of capitalism at its worst, but because the sale of mobile phone bandwidth, for example, brought cash in for a government at the expense of telecom companies, it was generally considered to be a good thing, even though there was every possibility that those companies would simply pass on the cost to their customers.

Spectrum of knowledge

The reason auctions are such powerful games is that they provide a mechanism to make the economic structure of the decision visible to the players. The way that other players bid gives information about the value they put on the item being auctioned. The use of game theory to maximise benefit from auctions was championed in the 1960s by economist William Vickrey. It was Vickrey who devised a type of auction (known unimaginatively as a Vickrey auction) that influenced the design of eBay.

As we have seen, eBay awards the win to the highest bidder, based on the value of the second-highest bid. A Vickrey auction also does this, but applies the approach to sealed-bid auctions. This auction subtype, as the name suggests, involves bids in sealed envelopes, which are not opened until the auction closes. In a traditional sealed-bid auction, the highest bidder has to pay whatever was in their envelope. This can lead to conservative bids, because bidders know that they will have to pay that much if they win. Conservative bids are not good from the seller's viewpoint. By ensuring the winner doesn't pay more than the second-highest bid, Vickrey's system encourages bidders to offer the maximum they can afford to pay. The mechanism extracts extra information from the bidders, even though that information is only ever known to the auction system.

It wouldn't be until 1994 that the game theory approach to auctions would really start to show benefits for the vendors. This was when the US Federal Communication Commission (FCC) first asked game theoreticians to be involved in the design of auctions to give access to telecommunications frequencies. The design of auctions has become so important that auction theory is sometimes labelled as a class of mathematics in its own right, though in reality it is just a subset of game theory.

Such spectrum auctions, when well-conceived, have brought in huge revenues for the relevant governments. The first raised $20 billion for the US government, while an equivalent auction in 2000 for the UK government's 3G mobile phone licences raised a remarkable £22.5 billion ($35 billion). The mechanisms behind these auctions are relatively simple in theory, if complex to run – however, if well designed, they prove extremely challenging games for the telecoms companies to play without paying a fair price for the licence.

Broadly speaking, most of these auctions are based on an approach known as a simultaneous multiple-round auction (or simultaneous ascending auction). As we saw in the example in Chapter 1, these usually involve many rounds of sealed bids for the different licences, which are auctioned simultaneous in groups. The bids are published at the end of each round and a minimum bid for the next round is calculated, which is usually the winning bid from the previous round plus a 5–10 per cent increment. The auction ends when a round has no new bids, the licence being awarded to the winner of the previous round.

No mechanism is perfect, however – and many auctions have been run with what could politely be called sub-optimal outcomes.

Takeover bid

The potential for poorly designed auctions to fail is made clear by the way that sales of the same slices of spectrum in different countries have brought in hugely varying amounts per potential phone user. When, for example, European countries were selling 3G licences in 2000, both the UK and Germany did extremely well, raising over €600 per person. By contrast, Austria, the Netherlands and Switzerland each raised less than a third of this rate, the latter coming in with a pitiful €20 per person.

What, then, could go wrong? As we have seen, the most obvious danger from using auctions is that they are susceptible to collaboration between bidders. If bidders talk to each other ahead of the game and all but one agree to hold back on a particular lot, then it can be bought at the minimum price set by the seller – the auction fails to do its job of establishing the market value of the lot. Because of this, high-value auctions such as spectrum auctions have meaningful punishments available if collusion is discovered. But this hasn't stopped would-be buyers from looking for game-playing approaches that enable them to share information without collaborating.

Often spectrum auctions are operated with multiple lots open simultaneously, so phone companies can bid on one or more from a wide range of lots at the same time. Without any prior collusion, as far as we are aware, bidders have managed to use the amounts they bid in the early stages to signal their intentions to other bidders.

Sometimes, the auction designers unwittingly give bidders a tool for achieving this communication. For example, in a 1999 German auction, in each round the bidders had to exceed the previous maximum bid by a minimum of 10 per cent. The idea was to avoid the auction taking too long as bidders made small increments in their bids from the previous round. In the first round, one company, Mannesman, bid 18.18 million deutschmarks per megahertz on one section of spectrum and 20 million deutschmarks on a second section. That strangely specific figure 18.18 million seemed designed to get the attention of their only significant rival for this bandwidth, T-Mobile. If you add 10 per cent to 18.18 million you get 19.998 million – to all intents and purposes 20 million. So, Mannesman seemed to be signalling to T-Mobile that beating Mannesman's 18.18 million bid with 20 million in the next round would enable both companies to get what they wanted at the same relatively low figure.

Was this collusion? Mannesman could argue honestly that no discussion had taken place between the companies, yet there is no doubt that information about intentions was passed from one to the other. It's hard to see how those running the auction could avoid this kind of tactic without changing the auction design, in ways we will discover in a moment. This indirect collusion mechanism effectively uses a carrot – one company offers the other the opportunity to do equally well in the next round. But it is also possible to use bid patterns as a stick, making it clear that retribution is threatened.

This was demonstrated in a 1990s US spectrum auction where two companies, US West and McLeod were bidding for a partial state licence (in the US, spectrum licences tend to cover relatively small parts of the country, rather than the country-wide licences typically sold in Europe). Each company seemed determined to get this Minnesota-based licence, lot number 378, pushing up the bidding. But then US West bid on two Iowa licences, which up until then McLeod had looked likely to get with little opposition.

Not only did US West outbid McLeod on the Iowa licences, it did so with the distinctive figures of $62,378 and $313,378. Most bids in the auction were in round thousands, so these values stood out, drawing attention to that number '378'. McLeod got the hint and stopped bidding on lot 378, after which they were given free rein to acquire the Iowa licences. Simply bidding an odd-looking value might have been enough, but US West's use of the '378' ending was particularly clear signalling. Arguably, here the authorities could have flagged up an explicit invitation to collude, but for whatever reason, they did not.

Bearing in mind the way that both these examples used strangely detailed figures to get their message across, one change that auction designers now often incorporate is a requirement that bids are rounded at a relatively high level, so that it would be very expensive to use bid values to communicate. (In the case of the Iowa licence, if bids were limited to round thousands, US West might have had to bid $378,000.) A second change has been to make bidding anonymous. The bids are still published at the end of each round, but the players are not told who made the specific bids, weakening any communication.

According to auctions expert Robert Leese, whom we met in Chapter 1, 'Quite a lot of effort has gone into spectrum auctions, to avoid collaboration. And that's done in general in two ways. First, through a strict application and scrutiny process, before bidders are accepted to take part in the auction. And secondly, very stiff penalties if any collaboration or collusion is found to have taken place, which would generally include removal from the auction completely.' In practice, though, as the examples above show, there has sometimes been reluctance to police spectrum auctions with any rigour.

Ruling the roost

All too often, when auctions have failed, the game rules were not carefully thought through. This is where the realisation than an auction is a game is so important. If the rules of a game are not clearly drawn up and shared, the result is ambiguity and conflict. Sometimes, the problem is that the rules are fine, but are not used properly. So, for example, many auction structures have a reserve price – a minimum that has to be reached if the sale is to go ahead. If this is set too low, as happened in the Swiss telecom auction mentioned on page 137, the result can be a very poor return. If it is set too high and is not reached, the sellers end up with egg on their faces.

Similarly, large-scale auctions with multiple simultaneous lots need strong rules on defaulting. There have been a number of telecom auctions in the past where the penalty for a company that changed its mind after winning a bid was very small compared with the price of licences. As a result, the companies best at gaming the system bid on more lots than they really wanted and chose the cream of the crop, discarding the rest. It is perfectly possible for there to be a big enough penalty for defaulting to prevent this, but the auction designer needs to make the rules watertight.

Punishments for misbehaving bidders must be carefully balanced too. In the 2000 3G auction for the Netherlands, the auction value was drastically reduced when one of the bidders made a spurious legal attack on another. Telefort threated to take Versatel to court, accusing the company of bidding up the prices on several licences with no intention to actually obtain the licence – supposedly an attempt by Versatel to damage their competitors by forcing them to pay over the odds. This was a claim made without evidence to back it up, yet the government did nothing to counter it, as a result of which Versatel withdrew from the auction.

With one less player, there were six companies vying for six licences, meaning that there was insufficient competition to reach a good price, leaving the Netherlands with a fraction of the expected income. Having the same number of parallel lots and existing companies limits competition as new entrants are less likely to attempt to take part. In a well-designed auction, there is usually at least one more licence than there are companies already in the market, and companies are only allowed to acquire one licence each.

Sealing facts

Although the most familiar auctions are those where bids are known during the auction, as we have seen there is an alternative approach, based on sealed bids. The biggest problem with a simple sealed-bid approach is that it cuts out the information-sharing that makes auctions so effective; this is mitigated, however, in the Vickrey style of auction described on page 135. Sealed-bid auctions can also be attractive as they tend to attract more participants than a conventional auction, as potential buyers reason there is no harm in putting in a relatively low bid which just may succeed.

A particularly sophisticated design to emerge from game theory is a hybrid approach, sometimes called Anglo-Dutch, where a conventional ascending auction is used until there are two bidders left, who are then invited to place a Vickrey sealed bid. The advantage here is that the bidders already have some information on the valuation of the lot, allowing them to set a more informed value in their final bid. It is also intended to prevent one rich participant totally dominating the game by repeatedly outbidding the competition – though in practice a conventional sealed-bid ending might work better in such circumstances.

Although eBay does not operate explicit sealed-bid auctions, the outcome can often feel like an Anglo-Dutch auction, because a highly contested lot will get a number of bids in the last few seconds of the sale. These act like sealed bids as there is no time for other bidders to respond to them.

Robert Leese notes the significance of information to the auction process: 'One of the decisions to make when designing an auction is the information policy. As the auction progresses, what information do the participants acquire? You can sometimes see this happening implicitly. If you go to a sale, you can get some indication of the demand there is at different price points as the auctioneer raises the asking price. For spectrum, at any rate, there tends to be information that is passed concerning the aggregate demand at different price points. So that would be different than in a sale, where the auctioneer is just going to take one bid wherever, whoever he spots putting their hand up first.'

This is one reason why these more sophisticated types of auction differ from the familiar auction room setting with an auctioneer simply looking for new bids. In a spectrum auction there are usually multiple items being sold and it is more like having an auctioneer ask 'Who will bid £100?', and then telling the whole room that there are ten people willing to pay £100, so who would be willing to give £200?

Leese points out that added sophistication also makes for greater complexity: 'Auction design gets more complicated when you have synergy between items being sold, as you do with spectrum licences. Very often, particular operators will want to acquire sufficiently many items in a geographical area or maybe items in a set of contiguous areas to build up their network coverage. Then, it isn't a straightforward case where the value of a set of licences is equal to the sum of the individual values; there's a complementarity effect as well. It becomes more important to bidders to know what the aggregate demand is for different items. Because then they can get some idea of what the pattern of demand is from the other participants and can adjust their own bids accordingly. And they can do this without knowing the identities of individual bidders. But revealing aggregate demand data is generally seen as a helpful way of assisting bidders in discovering what their best bidding tactics are.'

Gaming the gamers

One obvious consideration once game theory was brought to bear on auction design is whether the same theory could be used by bidders to finesse the outcome of auctions. We have seen gaming in bids that hide extra information, such as the US West bids of $62,378 and $313,378. But this isn't the only opportunity. Because of their underlying understanding of game theory, good auction designers look for opportunities to limit the ability of bidders to use strategy to gain an unfair advantage.

Sometimes the problem is not so much players using the system strategically to send information as a lack of clarity of what some of the information actually is. When, say, a house is being auctioned, bidders will usually have a reasonable idea of the value of the house based on sales of similar houses in the same area. But spectrum auctions introduce a considerable amount of uncertainty because they are often introducing a new technology, in which case phone companies have no real idea of how their customers will value access to that technology.

At the time of writing, licences for 5G, the next generation of mobile network, are being sold – and the telecom companies have little idea of what their customers are likely to be willing to pay for the extra functionality. The 5G service offers ultrafast connectivity, comparable to fibre broadband in speed, so forecasters might expect widespread uptake. But there is a limit to how much customers are prepared to pay – and savvy consumers are aware that the current highest speed network, 4G, is still not available everywhere, and can be very variable in quality; so it is entirely possible that they won't be prepared to spend a considerable amount on a technology that might not be fully delivering for a decade. As there is uncertainty at all of the mobile phone operators, they have even less chance than usual of knowing how their competitors will value the new licences. This means that auction designers have to construct a game for players who don't know themselves what their strategy should be.

Because information is so crucial to game theory, one strategy that is certainly deployed by the better players is to look beyond the information revealed by the auction to extra data that might give a bidder a competitive advantage over their rivals. In a house auction, for example, one bidder may know more than the others about, say, the other residents in a street and how they behave, which may make a house more – or less – valuable. Private information that is outside the auction system can often be exploited to hone strategy to a player's advantage, and this is likely to occur in auctions such as that for 5G.

Evil games

We have seen a number of opportunities for players in auctions to use a strategy that involves breaking (or at least stretching) the rules. One almost inevitable outcome of the study of game theory has been to experiment with games that have deliberately disadvantaging rules, which no one should sensibly play. RAND mathematician Martin Shubik came up with one of the best such games in the dollar auction. This involves auctioning off a dollar bill (if necessary, replace the dollar with a pound/euro/local divisible unit of currency of your choice). This is a standard forward auction with the highest bidder winning, but with the novel twist that the second-highest bidder *also* has to pay their bid, despite receiving nothing.

It is tempting for a player to initiate the process with, say, a one-cent bid. After all, who wouldn't trade one cent for a dollar? But another bidder will almost certainly come in at two cents. Now the first bidder faces paying one cent to receive nothing. Unless this is a two-player game, the chances are quite high that someone else will bid three cents or more, so the first bidder may hang on, as once someone else bids, they are off the hook. But if no one else bids, it is very much to the benefit of the first player to bid, say, three cents and get back on top.

This ramping-up process continues until someone has bid 99 cents. Why would anyone bid more? Because the second-highest bidder is currently due to spend 98 cents to get nothing. As a result, it's likely the 98-cent player will bid $1. They aren't losing anything by doing so, just breaking even. But now things get perverse. If the 99-cent bidder does nothing, he or she is due to lose 99 cents. But if they bid $1.01 – more than the value of the prize – they only lose one cent. It is worth carrying on beyond the value of the game to avoid making a greater loss.

In principle, the whole thing could spiral out of control, with bids rising for ever. In practice, there comes a point where most individuals decide to cut their losses. Once they see how it's going, players decide to give up the process and lose a relatively small amount rather than keep pushing up the bidding, so the bids rarely get very high as fewer and fewer players decide to pitch in. Nevertheless, when this game has been played for real, it has not been unusual for players to pay around $5 for a $1 bill and a cent less for nothing at all.

The dollar auction provides an effective model for some real-world situations, particularly queues where there is no indication of queuing time. Although some telephone queuing systems now tell you your position in the queue, which gives a feel for the rate of progress and distance to the front, many don't. In that case, the queuer is like the dollar auction player. If the queue takes too long, at some point, most will give up and 'pay' their time spent queuing for no return.

Theme park queues are sometimes designed deliberately to conceal the information that might enable queuers to decide when to cut their losses, hiding how long the queue really is by continuing it unseen inside a maze-like structure. And for that matter, a dollar auction queue may be virtual. Think, for instance, of waiting for a bus. How long do you wait at the stop before you give up and walk, throwing away time for nothing? On the bus route from the railway station to my home, I have sometimes played a more constructive variant of this game. If I walk along the bus route for fifteen minutes, I can get the same bus at a considerably cheaper fare. But if I do so, I may miss a bus and have to wait longer. At one time, this was a high-risk strategy, but now that it's possible to track buses on the internet, it can be a profitable game to play.

A variant of the dollar auction is a two-player version where the players are allowed to discuss strategy. In this case it turns into a variant of the ultimatum game (see page 95). I might say to my opponent, *let me win with a one-cent bid. I will then give you 49 cents*. We both get the same profit with no risk, other than the possibility that I will renege. Players who trust each other can benefit most from this strategy. If, on the other hand, I bid one cent and only offer the other player five cents to stay out of the game, they may well respond similarly to the reaction to a low offer in the ultimatum game. The equivalent here would be to outbid me to punish that feeble offer. If I had a good grasp of game theory, I might then ask them for, say, a payment of 40 cents not to bid again, so that they feel vindicated in inflicting some revenge, but we both still benefit.

Is the brain playing games?

Beyond the technical sphere of auctions, game theory can seem like abstract mathematics, removed from the real world, because it works on the assumption that players are rational and make the best choice given the available data – a picture that can seem very different from the reality of human action. However, there is evidence that at least some aspects of the way the brain functions at the level of the interaction of neurons reflect the kind of logical processing underlying game theory.*

In normal life, most people don't sit down and calculate a matrix of outcomes when they make a decision: the very process of working out the outcomes can make the brain feel frazzled. But that doesn't mean that there isn't some mental processing of this sort going on. We can see this particularly in the way that the brain appears to use Bayes' theorem. This is a powerful tool in probability that often seems counterintuitive, yet it is similar in approach to the way that networks of neurons interact.

---

* Neurons are interconnected cells in the brain that effectively undertake calculations by the way that the cells are connected and pass on a signal when their input reaches a certain trigger level.

Bayes' theorem is a mechanism for translating probabilities we already know into probabilities we actually want to know. A particularly timely example, given this book was written during the Covid-19 pandemic, is the effectiveness of medical tests. At the time of writing, there are two main types of test. PCR (polymerase chain reaction) tests are the best available, but are expensive and can take a couple of days to process. The alternative lateral flow tests can be turned round in under an hour and are a lot cheaper, but are less reliable.

All tests have two areas where probability rears its head. How often will the test say you don't have the disease when you do (false negatives), and how often will it say you do have the disease when you don't (false positives). The ability of the test to avoid false negatives is called its sensitivity, while its ability to avoid false positives is its specificity. As the lateral flow tests are the less reliable of the two types, it's particularly important to get to grips with the probability of a result being meaningful when one of these is used.

On a recent study, the best current lateral flow tests had a specificity of 99.68 per cent and a sensitivity of 76.8 per cent when operated by a professional lab, which dropped to 57.5 per cent when performed by home testing services. That lower percentage is the more relevant one, as the whole point of using lateral flow tests is to avoid the delays incurred by using a lab – for the rest of this example, we'll use the lower sensitivity figure. The specificity tells us that the chance of getting a positive result if you don't have Covid is just 0.32 per cent, while the sensitivity tells us that the chance of getting a negative result if you do have Covid is 42.5 per cent. But these are not necessarily providing the information we really want.

Let's take the example of false positives. The specificity

is the chance of getting a positive result if you don't have Covid. But what I'd really like to know is the chance I've not got Covid if I get a positive result. Bayes' theorem is a mechanism to convert one probability into the other – but we need a couple more pieces of information. One is the infection rate and the other is how many people take the test. At the time of writing, the national average infection rate of Covid is around 40 in 100,000 of the population, while the number of lateral flow tests being taken is around 1 million a day. So, of the million people tested, perhaps 400 have the disease. Of these, around 230 will be detected by home testing. Similarly, of the 999,600 who don't have the disease, 0.32 per cent will get false positives, making around 3,200.

We can say that 3,200 of the positive test results are false while 230 are real. So, there is a $^{230}/_{3,430}$ chance if you get a positive result that you have the disease – about one in 14.9, or a 6.7 per cent probability. In practice, things are more complex than this as the disease isn't evenly spread across the country. These figures only apply in an average location – if you were in a location that didn't have the average number of infections, you would need to adjust that 400-in-a-million figure accordingly.

Most of us are more worried about false negatives than false positives. A false positive means that you might have to isolate for a period unnecessarily, but a false negative means you could go about your business while infectious, or could develop a serious illness. The sensitivity, 57.5 per cent, is the chance you will get a positive result if you have the disease, so in 42.5 per cent of cases you will get a negative result if you have the disease. But what is the chance you have got the disease if you have a negative result? With that same proviso about the distribution of the disease around the country, of those getting a negative result, 996,400 won't have the disease while 170 will. So the chances you do have the disease after a negative result are $^{170}/_{996,400}$ – a 0.017 per cent chance.

We now have a better understanding of the test results. If I get a positive result, while I should self-isolate, I shouldn't panic as there's a 93.3 per cent chance that I don't have the disease. However, Bayesian statistics requires us to update the outcome if we have extra prior information. If, for example, I have symptoms I am not part of the general population but of the smaller population who have Covid symptoms. Here, far more than 40 in 100,000 have the disease. If I get a negative result, although false negatives will be rare in the population as a whole, because of the relative inaccuracy of the test, it is worth having a second test before assuming the negative result is correct.

Game theory and reality

When we apply Bayes' theorem to a real-life situation it can seem counterintuitive. Doing the maths isn't natural to us. But our brains appear to have circuitry that carries out this kind of mechanism without any need to crunch the numbers. For example, studies of the way our movements are controlled suggest that the brain combines data from experience with sensory inputs and ideas of how reliable the data is to produce the resultant movement. Like more conventional aspects of game theory, Bayes' theorem gives us a chance to make a better decision.

Although most economists never got the hang of game theory, it remains widely used in other fields. A search of the literature throws up thousands of papers each year, employing game theory in a startling range of applications. Looking at papers published in 2019–2020, particularly apt is one that uses game theory to decide which is the more effective, quarantining as a preventative measure or isolating those with an infectious disease. (The recommendation was to do both when the prevalence is high, but quarantining has more impact than isolation when prevalence is low.)

Other topics included management software strategies for the internet of things, supply-chain modelling under labour restraints, designing green incentives, promoting prefabricated building projects, waste management, scheduling drone charging, motorway lane changing, risk assessment of third-party damage to oil and gas pipelines, a model of gut bacteria interactions in infants, and credit card fraud detection. The methods of game theory may not be at the cutting edge of mathematics anymore, but the theory provides a powerful method that can be widely applied in modelling studies.

For me, although I am unlikely to make use of game theory in everyday decision-making, it still has a huge benefit. Game theory helps us get a better feel for how we deal with dilemmas and interact with others in everyday life, uncovering our values. Just as the models of physics simplify the physical world but still work, so the tables and trees of game theory provide simplistic models of human decision-making, negotiation and competition that remain useful.

We are no more likely to draw up a game theory table when choosing which ice cream to buy in the supermarket than we are to work out the mathematics of the Newtonian dynamics involved when we throw a ball, but each of these models helps us understand what is happening better. And surely this is of particular importance when humans interact and hope to make the best possible decisions.

FURTHER READING

AlphaGo article: 'Mastering the game of Go without human knowledge', *Nature* 550, 354–359 (2017).

Auction design detail: *Discovering Prices: Auction Design in Markets with Complex Constraints*, Paul Milgrom (Columbia University Press, 2017).

Auction design flaws: 'What Really Matters in Auction Design', Paul Klemperer, *Journal of Economic Perspectives* 16:1, Winter 2002, 169–189. Interesting paper on what can go wrong with auctions.

Bayesian brains: 'Are our brains Bayesian', Robert Bain, *Significance* (Royal Statistical Society), August 2016, 14–19. Very readable article on the use of Bayesian inference by the human brain.

Game theory detail: *Theory of Games and Economic Behavior*, John von Neumann and Oskar Morgenstern (Princeton University Press, 1953). The game theory bible. Very technical (and long).

Game theory graphic guide: *Introducing Game Theory*, Ivan Pastine, Tuvana Pastine, Tom Humberstone (Icon Books, 2017). Visual guide to game theory concentrating more on what it teaches us than the theory itself.

Game theory practicalities: *The Compleat Strategyst: Being a Primer on the Theory of Games of Strategy*, J.D. Williams (Dover Publications, 1986). Rather odd introduction to game theory with distinctly dated (1950s) humour. Goes into some depth on solutions for different types of game.

John Nash: *A Beautiful Mind*, Sylvia Nasar (Faber & Faber, 1998). The source of the film of the same name: a sympathetic biography, better on the personal aspects of Nash's life than the mathematical content of his work.

John von Neumann: *John von Neumann: The Scientific Genius Who Pioneered the Modern Computer, Game Theory, Nuclear Deterrence and Much More*, Norman Macrae (Pantheon Books, 1992). Interesting if distinctly quirky biography of von Neumann.

Limitations of economics: *Economyths: 11 Ways Economics Gets It Wrong*, David Orrell (Icon Books, 2017). Excellent assessment of the failings of economics as a mathematical science.

INDEX

BEHAVIOURAL ECONOMICS

Psychology, neuroscience, and the human side of economics

DAVID ORRELL

For centuries, economics was dominated by the idea that we are rational individuals who optimise our own 'utility'. But it turns out the reality is a lot messier. We don't really know what our utility is, and we care about people other than ourselves. We are susceptible to 'nudges'. And far from being perfectly rational, we are prone to 'cognitive biases', with complex effects on our decision-making. David Orrell explores the findings from psychology and neuroscience that are shaking up economics – and assesses the lofty claims made for this most image-conscious of disciplines.

ISBN 9781785786440 (paperback) / 9781785786457 (ebook)